T0364171

MULTIMEDIA GROUP COMMUNICATION

MULTIMEDIA GROUP COMMUNICATION

PUSH-TO-TALK OVER CELLULAR, PRESENCE AND LIST MANAGEMENT CONCEPTS AND APPLICATIONS

Andrew Rebeiro-Hargrave

Nokia Siemens Networks, Africa

David Viamonte Solé

Genaker, Spain

John Wiley & Sons, Ltd

Other Wiley Editorial Offices

John Wiley & Sons Inc., 111 River Street, Hoboken, NJ 07030, USA

Jossey-Bass, 989 Market Street, San Francisco, CA 94103–1741, USA

Wiley-VCH Verlag GmbH, Boschstr. 12, D-69469 Weinheim, Germany

John Wiley & Sons Australia Ltd, 42 McDougall Street, Milton, Queensland 4064, Australia

John Wiley & Sons (Asia) Pte Ltd, 2 Clementi Loop #02-01, Jin Xing Distripark, Singapore 129809

John Wiley & Sons Canada Ltd, 6045 Freemont Blvd, Mississauga, ONT, L5R 4J3, Canada

Wiley also publishes its books in a variety of electronic formats. Some content that appears in print may not be available in electronic books.

Library of Congress Cataloging-in-Publication Data

Rebeiro–Hargrave, Andrew.
    Multimedia group communication : Push-to-Talk over cellular, presence, and list management concepts and applications / Andrew Rebeiro–Hargrave, David Viamonte Solé.
        p. cm.
    Includes bibliographical references and index.
    ISBN 978-0-470-05853-4 (cloth)
1. Multimedia communications. 2. Telephone conferencing. 3. Internet telephony. I. Viamonte Solé, David. II. Title.
    TK5105.15.R43   2008
    621.3845′6 – dc22

British Library Cataloguing in Publication Data

A catalogue record for this book is available from the British Library

ISBN 978-0-470-05853-4 (HB)

Typeset by SNP Best-set Typesetter Ltd., Hong Kong
Printed and bound in Great Britain by Antony Rowe Ltd, Chippenham, England.

Table of Contents

Foreword

Here is a book with pioneers in action.

When the world wide web was being shaped, few observers realised what was emerging.

This book documents an exciting step which reaches beyond classic telephone services. It is too early to see which inventions will last, but it is exciting to observe this innovation in action. Multimedia group communication, as described in this book, is an inspiring invention stretching the envelope of new services based on internet technologies. Some of the brightest minds in the industry saw how group communication capabilities seen previously in closed enterprise services based on trunking radio could be brought to the mainstream consumer market based on mass market terminals and data connectivity using GSM/GPRS infrastructure.

Combining session management and streaming technologies, a service is provided where any member of a group may take the floor and speak to others. One can think of this as an enhanced chat session. Rather than writing a comment on a shared chat board, here we exchange observations and comments in a shared voice channel. This is more convenient, in particular, when on the move.

Group communication like this enables teams spread across physical locations to share comments, helping them maintain a view of the status of other team members. It also enables casual peer groups at leisure to, keep in touch when apart. Establishing such a shared context would be a tedious exercise using classic voice calls, conference calls or exchange of text messages.

Group communication changes the way we think about our contacts. They no longer look like a plain alphabetical list of entries. With group communication we see individuals in the context of the groups where they are members. Presenting such structured contacts is an essential part of implementing a group communication service.

While considering initiating a session or during a session, we are interested to observe the actual availability status of members of our group. Hence presence becomes an active on-line attribute of high relevance.

Many of us would be surprised to realise that the authors of the book do not describe a conceptual theoretical model, but a real implemented service. The latest mobile terminals have a group communication feature (often known as push to talk) already included and many network operators already have deployed the infrastructure needed to offer such a service.

Understanding this new communication paradigm helps us to realise how internet based technologies will enrich communication.

Petri Pöyhönen
Head of Converged Internet Connectivity Business Line
Nokia Siemens Networks

Preface

Multimedia Group Communications encapsulates three enablers – Push-to-Talk over Cellular (PoC) service, XML Document Service (XDMS), and Presence with SIMPLE service – that have been standardized by Open Mobile Alliance (OMA). These services combine to allow mobile users to create and store their own groups, and communicate with the group members in real time.

Group communications and Push-to-Talk over Cellular in particular is a topic of interest in the mobile industry within the last ten to twelve years. After the success of the Push-to-Talk service in the US during the mid-nineties, operators worldwide started to look at *walkie-talkie over cellular* service with attention. With this interest in mind, a group of leading mobile infrastructure and handset vendors developed the *Industry Consortium* for PoC. Regardless of the effort, this work did not move on to the commercial stage. Rather, different non-compatible PoC solutions were commercially deployed in several countries around the world in the early 2000s (Germany, New Zealand, Japan, Mexico, Sweden, Spain).

Pre-standard PoC solutions have made an important contribution in bringing the cellular walkie-talkie user experience to the market. However, at some point in time it became evident that, for a community-based service, it is crucial to ensure interoperability across handsets (*I can PoC any other user, regardless of the handset they use*), and across networks (*I can PoC any other user, regardless of the network operator they are subscribed to*). A common playground (a standard) is required to ensure such degree of interoperability, as a key enabler of real service take-up and success.

With this idea in mind, the Open Mobile Alliance started standardization activities around PoC, Presence and Group Management (later on renamed as *XML Document Management*) during 2003. The results of that work cristalized in 2006, when the three enablers were officially approved as a first step to let device and infrastructure providers develop standards-based products.

At the time of writing (September 2007), there are interesting signs in the market that operators are progressively migrating their pre-standard deployments towards an OMA compliant infrastructure. The first OMA PoC / Presence / XDM devices are already available, with new models and brands progresively incorporating PoC during 2008. Operators are now again looking at PoC and group communications with attention, possibly focusing on the corporate sector as a first step, as a critical mass of handsets becomes available prior to a residential PoC launch.

Interestingly, PoC is the first one of a new paradigm of SIP-based services to be deployed in a large scale. Effectively, Push-to-Talk over Cellular is the first service providing a real *group communication* experience (it is very easy to build up and communicate between groups of users with PoC), and it is the first service that delivers a *real-time communications* experience over the *packet-switched* cellular network (before PoC, most real-time services were run on top of the circuit-switched infrastructure). It comes as no surprise that these two paradigms represent a new step in the communications industry. In fact, these concepts are the foundation for future, innovative services that are already in the standardization pipeline or about to reach commercial status, such as SIP-based messaging or SIP-based multimedia conferencing.

Yet, it is understood that the deployment of these new communication paradigms introduced by PoC is not a trivial task. It requires developing new sets of skills and solutions across handset and infrastructure vendors, network operators and service providers. Furthermore, users need to get used to the service and understand the value it brings to their everyday life, both in the corporate and in the consumer sectors. It is our very modest aim with this book to try to help all these players make their move towards this new communications concept.

This book was written to provide detailed insights about the new multimedia group communication experience in general, and PoC, Presence and XDM enablers in particular – the concepts, architecture, protocols, application and future orientation. Its intended audience ranges from marketing managers, research and development engineers, and test engineers to university students. The book is written in a manner that allows readers to choose the level of knowledge they need and the depth in understanding they desire to achieve about multimedia group communication. The book is also very suitable as a reference. Each chapter can read as individual source and references are given for further study. We briefly describe the book structure below.

Part I defines the concepts and gives a detailed overview of the system architectures and entities that, when combined, support the group communication service. Chapter 1 provides an overview on the main concepts associated with group communication. Chapters 2–4 provide details for each enabler –such as their respective architecture and associated protocols and specific features.

Part II gives a more practical viewpoint of group communication and focuses on applications. Chapter 5 cover deployment issues – integration with IMS, identity management, PoC charging and radio issues. Chapter 6 gives a step by step example of the PoC service at the protocol level, detailing the procedures taken at every entity and paying special attention to signalling flows.

Part III alludes to evolution of multimedia group communication and points out future opportunities. Chapter 7 focuses on the present, and combines PoC and Presence with current Value Added Services. Chapter 8 turns to the future and discusses new concepts introduced by OMA PoC2. Chapter 9 finalises describing the architectural evolution of OMA enablers: PoC2, XDMS 2, Presence 2 and SIMPLE instant messaging.

The original idea of this book was born at OMA Test Fests 2005–2006. Whilst working through countless interoperability test cases and endless cups of coffee with industry vendors, it was thought prudent to simplify the good work put together by the OMA forum. From that point on, we as authors applied our observations on PoC, XDMS and Presence and compiled this publication.

We want to thank all colleagues in our companies and in the industry that had ideas and the ability to create multimedia group communication technology in the mobile domain. We also thank the people OMA who had the patience to bring this technology to life through the process of standardization.

Acknowledgements

The authors of the book extend their thanks to contributors working in the Open Mobile Alliance for their great efforts in creating the specifications and related protocols. We especially appreciate the support of Mark Hammond and enthusiasm and patience of Sarah Hinton, our editor at John Wiley & Sons, through the whole creation process.

Andrew would like to thank colleagues at Nokia Siemens Networks, Nokia Oy, organizers of OMA Test Fests (9-15), and vendors that participated at the Test Fests. In particular, I acknowledge the guidance and assistance I received from Juha Kallionen, Juha-Matti Liukkonen, Aki Koivisto, Tapio Alanen, Jarmo Lindberg, Simo Suominen, Eija Junnila, Antti Toivanen, Roman Smirak, Martin Hynar, Silvestr Peknik, Jyri Sarha, Saku Oja, Tapio Paavonen and Vladimir Mijatovic. On a more personal level, I thank Paula and more recently Luna, who endured my efforts as this publication was constructed.

David would like to thank Belen and Clara (my wife and daughter) for their endless understanding, support and enlightening through the project. A warm thank you to all colleagues at Nokia, Vodafone and Genaker, from whom I have borrowed all the learnings necessary to participate in this project: Luis Khamashta, Frank Timphus, Haris Zisimopoulos, Vidhya Gholkar, Marco Stura, Martin Guntermann, Heiko Gerlach, Barry Gallagher, Manuela Salonia, Richard Powell, Michael Hillier, Mike Prince, Alfonso Hidalgo, Pascal Maugeri, Javier Rodríguez, Jordi Pratsevall and many more. Thanks to Jordi Guerrero and Miquel Teixidor, who supported this project from day one. Thanks to Josep Paradells as the beginner of it all, to Anna Calveras, Neil Jackson, Jürgen Tibes, Luis López and Fraser King. Finally, thanks to Juan Pablo Calvo for his enthusiasm and teachings about PoC.

We thank Genaker for the authorization to use its image library – which is in turn inspired in OpenCipArt – in some of the figures of the book.

The authors welcome any comments and suggestions for the improvements or changes that could be used to enhance future editions of this book. Our email addresses are:

andrew.rebeiro-hargrave@nsn.com
david.viamonte@genaker.net

Abbreviations

3GPP	3rd Generation Partnership Project
3GPP2	3rd Generation Partnership Project 2
ACL	Access Control List
ACR	ACcounting Request
AKA	Authentication and Key Agreement
AMR	Adaptive Multi Rate
AoR	Address of Record
APN	Access Point Name
APP	Application defined RTCP packet
ARPU	Average Revenue Per User
AS	Application Server
AUID	Application Unique ID
AVP	Attribute Value Pair
B2BUA	Back to Back User Agent
BGCF	Break-Out Gateway Control Function
BSC	Base Station Controller
BTS	Base Transceiver Station
BW	Band Width
CCA	Credit Control Answer
CCF	Charging Collection Function
CCR	Credit Control Request
CDF	Charging Data Function
CDMA	Code Division Multiple Access
CDR	Charging Data Record
CGF	Charging Gateway Function
CID	Content ID
CLIR	Calling Line Identification Restriction
CMR	Codec Mode Request
CNAME	Canonical name
CIPID	Contact Information in Presence Information Data Format
CPF	Controlling PoC Function
CS	Circuit Switched
DCCA	Diameter Credit Control Application
DM	Device Management
DMS	Device Management Server

DND	Do Not Disturb
DNS	Domain Name Service
DTMF	Dual Tone Multi-Frequency
EBCF	Event Based Charging Function
ECUR	Event Charging with Unit Reservation
EDGE	Enhanced Data Rates for the GSM Evolution
ENUM	Telephone NUmber Mapping
EPA	Event Publication Agent
ESC	Event State Compositor
EVRC	Enhanced Variable Rate Codec
FDCFO	Full Duplex Call Follow-On
FQDN	Fully Qualified Domain Name
GA	Group Advertisement
GAA	General Authentication Architecture
GEOPRIV	GEOgraphic Location and PRIVacy (IETF WG)
GERAN	GSM/EDGE Radio Access Network
GGSN	Gateway GPRS Support Node
GML	Geographic Mark-up Language
GPRS	General Packet Radio Service
GPS	Global Positioning System
GSM	Global System for Mobile Communications
GSU	Granted Service Units
GUI	Graphical User Interface
HDVC	Half Duplex Voice Chat
HLR	Home Location Register
HSDPA	High Speed Downlink Packet Access
HSS	Home Subscriber Server
HTTP	HyperText Transfer Protocal
HTTPS	Secure HyperText Transport Protocol (HTTP over TLS)
IAB	Incoming Personal Alert Barring
IANA	Internet Assigned Numbers Authority
IEC	Inmediate Event Charging
IETF	Internet Engineering Task Force
iFC	initial Filter Criteria
IM	Instant Messaging
IMPI	IP Multimedia (IMS) PrIvate User Identity
IMPS	Instant Messaging and Presence Service (aka Wireless Village)
IMPU	IP Multimedia (IMS) PUblic User Identity
IMS	IP Multimedia Subsystem
IMSI	International Mobile Subscriber Identifier
IN	Intelligent Network
IP	Internet Protocol
IP-CAN	IP – Connectivity Access Network
IPA	Instant Personal Alert
IPIIM	Invited Party Identity Information Mode
ISB	Incoming Session Barring

ISC	IMS Service Control Interface
ISIM	IMS Subscriber Identity Module
IVR	Interactive Voice Response
LBS	Location Based Services
LI	Lawful Interception
MAO	Manual Answer Override
MBCP	Media Burst Control Protocol
MCC	Mobile Country Code
MIDP	Mobile Information Device Profile
MIME	Multipurpose Internet Mail Extensions
MMD	MultiMedia Domain
MMS	Multimedia Messaging Service
MNC	Mobile Network Code
MO	Management Object
MRF(C/P)	Media Resource Function Controller/Processor
MSC	Mobile Services Switching Centre
MSIN	Mobile Subscriber Identity Number
MSISDN	Mobile Subscriber Integrated Services Digital Network Number
MSRP	Message Session Relay Protocol
MWS	Mobile Web services
NACC	Network Assisted Cell Change
NAME	User Name SDES Item
NNI	Network-to-Network Interface
NTP	Network Time Protocol
O-CTF	OMA Charging Trigger Function
OCS	Online Charging System
OMA	Open Mobile Alliance
OMNA	OMA Naming Authority
OTAP	Over the Air Provisioning
P2HDVC	PoC to Half Duplex Voice Chat
P2P	Peer to Peer
P2T	Push-to-Talk
P2VIM	PoC to Voice IM
PCRF	Policy and Charging Rules Function
PCU	Packet Control Unit
PDA	Personal Digital Assistant
PDF	Policy Decision Function
PDN	Packet Data Network
PDP	Packet Data Protocol
PEA	Presence External Agent
PEP	Presence Enabled Phonebook
PIDF	Presence Information Data Format
PLMN	Public Land Mobile Network
PMR	Private Mobile Radio
PNA	Presence Network Agent
PoC V1.0	Push-to-Talk over Cellular, Version 1

PoC V2.0	Push-to-Talk over Cellular, Version 2
POI	Point of Interest
PPF	Participating PoC Function
PS	Presence Server
PS	Packet Switched
PSI	Packet System Information
PSI	Public Service Identity
PSL	Presence Subscription List
PSTN	Public Switched Telephone Network
PT	Payload Type
PTT	Push-to-Talk
PUA	Presence User Agent
P-CSCF	Proxy Call State Control Function
QoE	Quality of Experience
QoS	Quality of Service
RAN	Radio Access Network
RFC	Request For Comments (IETF specification)
RLMI	Resource List Meta Information
RLS	Resource List Server
RPID	Rich Presence Information Data
RR	Receiver Report
RRC	Radio Resource Control
RTCP	RTP Control Protocol
RTP	Real-time Transport Protocol
R-UIM	Removable User Identity Module
SCR	Static Conformance Requirement
SDES	Source Description RTCP Packet
SDP	Service Delivery Platform
SDP	Session Description Protocol
SGSN	Serving GPRS Support Node
SIM	Subscriber Identity Module
SIMPLE	SIP Instant Message and Presence Leveraging Extensions
SIP	Session Initiation Protocol
SMPP	Short Message Peer-to-Peer Protocol
SMS	Short Messaging Service
SMTP	Simple Mail Transfer Protocol
SR	Sender Report
SSL	Secure Socket Layer
SSRC	Synchronization source
SSS	Simultaneous PoC Session Support
S-CSCF	Serving Call State Control Function
TBCP	Talk Burst Control Protocol
TBF	Temporary Block Flow
TCP	Transport Control Protocol
TETRA	Terrestrial Trunked Radio
TLS	Transport Layer Security

TSL	Time Slot
UA	User Agent
UAC	User Agent Client
UAS	User Agent Server
UDP	User Datagram Protocol
UE	User Equipment
UI	User Interface
UIM	User Identity Module
URI	Uniform Resource Identifier
URL	Uniform Resource Locator
USIM	UMTS Subscriber Identity Module
UMTS	Universal Mobile Telecommunications System
UTRAN	UMTS Radio Access Network
VAS	Value Added Service
VoIP	Voice over IP
VXML	Voice XML
WAP	Wireless Application Protocol
WCDMA	Wideband Code Division Multiple Access
WG	Working Group
WLAN	Wireless LAN
WPS	Wireless Priority Service
WV	Wireless Village
XCAP	XML Configuration Access Protocol
XDM	XML Document Management
XDMC	XML Document Management Client
XDMS	XML Document Management Server
XML	Extensible Markup Language
XUI	XCAP User Identity

1

Group Communication Concepts

1.1 Introduction

Group communication consists of two fundamental components – static data and session data. Static data is associated with persistent lists. For example, a service provider is associated with a list of subscribers; a freight company is associated with a list of drivers; a school is associated with a list of students; an end user is associated with a list of contacts on a mobile phone. Session data involves transient interactions between the listed members. For example, in a freight company a dispatcher interacts with other drivers when giving delivery instructions; a student interacts with other student when making a presentation during a conference; the end user interacts with one or more or subscribers when making a call. When one thinks about it, the combinations of listed members and sessions between members are actually quite endless.

This chapter introduces the reader to basic group communication concepts. Section 1.2 explains that role definition and administration are the first requirements for group communication. Section 1.3 defines a group session in terms of group function execution and gives a series of examples. Section 1.4 outlines the basic Internet protocols and signalling concepts required for implementation of OMA enablers – Push-to-talk over Cellular (PoC), XML Document Management (XDMS), and Presence with SIMPLE.

1.2 Group Communication Roles

Group communication involves the setup of a multimedia session between two or more users, in other words a list of users. A list is normally ordered in terms of hierarchy or the importance of its elements and there are different assigned roles for the arranged items or members. A role defines the actions and activities expected by the group and members. This section concentrates on two roles: the service provider and the end user [1].

1.2.1 Service Provider

The service provider assumes the dominant role for multimedia group communication. The service provider owns or leases the infrastructure and has invested in:

Multimedia Group Communication. Andrew Rebeiro-Hargrave and David Viamonte Solé
© 2008 John Wiley & Sons, Ltd.

- SIP/IP Core and Application Servers that allow subscribers to form groups and conduct peer to peer multimedia communication – that is to enable users to '*set up and talk in their own groups*';
- A Radio Access Network to allow subscribers to quickly form mobile groups and '*talk in real time*', in which the participants may not be physically visible to each other.

To ensure a successful group communication service, the service provider needs to ensure clear administrative strategies:

- A marketing strategy that constantly informs their subscribers of the general concepts of group communication, how to relate mobile group talk to their lives, and how to administrate the experience of talking and listening to other group members;
- An attractive price strategy for group communication. Since group communication is predominately a listening/receiving service – *one person speaks/sends the other five participants' listen/receive* – the aggregate price for listening or receiving data should always be low;
- An excellent quality of service strategy, in which the call setup and the round trip time of voice from the speaker party to the group listeners appears to be almost instant. The quality of service should mimic a professional walkie talkie experience.

1.2.2 End Users

The end user's role is to subscribe to the service provider's group communication service. The subscriber uses client(s) to convey their group communication desires. For mobile group communication, the client is embedded on the mobile phone. For stationary group communication, the client can be embedded on a PC or any other fixed device to perform the enabling operations. The PoC user is the same as a PoC Subscriber and PoC users become participants when engaged in a PoC session.

We will consider two types of PoC users for group communication:

- Basic Users
- Professional Users.

1.2.2.1 Basic PoC Users

Basic PoC users are the service provider's subscribers who purchased the service to form group talk sessions for entertainment, hobbies, college events, and to coordinate small businesses. Basic PoC users may form a PoC community. Basic PoC users use Access Control Lists (ACL) to define their level of interaction with other PoC users. An Access Control *Reject List* contains rules that restrict other PoC users who may try to establish a PoC session with the PoC user that owns the list. An Access Control *Accept List* allows other PoC users to talk to the PoC user instantly. Access Control Lists are administered using the XDMS service and are stored in the PoC XDM server. These will be described in Chapter 3.

1.2.2.2 Professional PoC Users

Professional PoC users are the service provider's enterprise subscribers, such as government departments, utilities and freight companies, who purchase the service to coordinate their work force. These organizations may allocate PoC Group Administrator rights to define, delete or modify PoC group memberships. Professional PoC users can also be served by a Value Added Service Provider who specializes in providing group communication to niche industries.

Professional PoC users apply role definitions to control communication. As an example, when the 1-to-Many-to-1 PoC communication feature is used, a Distinguished Participant can send and receive media to/from all Ordinary Participants, whilst Ordinary Participants can only send and receive media to/from one Distinguished Participant. *A Distinguished Participant is similar to a dispatcher in a police force who can talk to many different groups. The Ordinary Participant is similar to a policeman, who can only talk to the one nominated dispatcher.* In addition to defining who is who, the dispatcher may be enabled to apply priority talk rights to certain individual participants. A person with pre-emptive priority can interrupt the PoC speech of another user with lower priority. For example, if a worker is talking in a group session – *holding the floor* – a manager can interrupt before the worker has finished his sentence – *taking the floor* – and give commands to the listening participants. Formal OMA definition of roles for PoC group communication is given in Table 1.1.

Figure 1.1 shows example roles involved in mobile group communication. The Service Provider infrastructure simultaneously supports different types of PoC Groups and related

Table 1.1 Push-to-Talk over Cellular main communication roles

Terms	OMA Definitions
PoC Administrator	An entity that creates and maintains relevant aspects of PoC service for a specific PoC subscriber or group of PoC subscribers. The PoC service provider is the default PoC administrator. *In this book a service provider can be a cellular network operator such as pocservice@ Wirelessfuture.com*
PoC Group Administrator	A person(s) or entity that has the authority to define, delete or modify PoC group memberships. The PoC service provider has group administrator rights by default. The service provider can allow Value Added Service Providers and individual organizations to create, modify and delete their own groups
PoC Subscriber	One whose service subscription includes the PoC service. A PoC subscriber can be a PoC user such as *John@Wirelessfuture.com*
PoC User	PoC user can be the same person as a PoC subscriber. A PoC user uses the PoC features through the User Equipment
Participant	PoC user in a PoC session
Distinguished Participant	A participant in a 1-many-1 PoC group session that sends media to all Ordinary Participants, and that receives media from any Ordinary Participant
Ordinary Participant	A participant in a 1-many-1 PoC group session that is only able to send and receive media to and from the Distinguished Participant

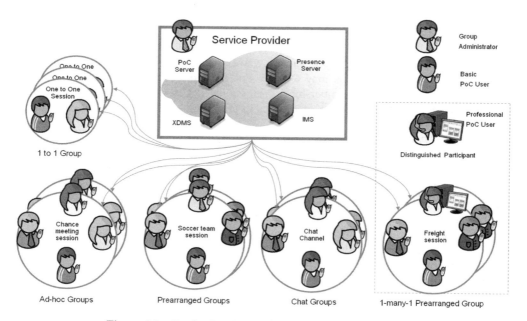

Figure 1.1   Service Provider, PoC Users, Groups and Sessions

sessions. The Service Provider is the root Group Administrator and allows the PoC community to form the different group types. The Service Provider comprises four services – SIP/IP Core (referred to as IMS), PoC Service, Presence Service and Group List Management Service (XDMS). There are four session types – 1 to 1, Ad-hoc, Pre-arranged, and Chat. The Chat session can be further subdivided into Open Chat and Restricted Chat session. PoC users are participants in a PoC session. For Professional PoC, the participants may be subdivided into Distinguished Participant (PoC Dispatcher) and Ordinary Participants.

1.3 Mobile Group Communication Use Cases

It is the service provider's task to market group communication use cases. For ease of explanation, PoC group communication can be simplified into the following:

- A PoC group is a predefined set of users that can be called:
  - ○ Users listed in contacts on User Equipment
  - ○ Users listed in documents stored on the network
- A PoC session is the real time connection between a set of users:
  - ○ The set of users can be invited to the session by another PoC user
  - ○ The set of users can join in to the session by their own free will – this is similar to joining an Internet chat room channel.

The separation between PoC group and PoC session involves the XML Document Management (XDMS) service. XDMS allows end users to create lists (group documents) to predefine the membership of a group and store these documents on a network server. The documents can be downloaded by the end user for list modification (*I want to remove Paul from my list*), when they upgrade their mobile devices (*I have just bought a new mobile phone and do not want to create the groups again*), or when they move from device to device (*I want to access my group from both my mobile and desktop PC*). The document can also be downloaded by other services such as a Presence server, so the end user may publish his or her availability to PoC to those on the group list. The service provider can capture loyalty from the PoC community by storing group documents on their network.

The PoC session is the execution of the group function. When an individual wants to form a group session there are two modes of execution: the dial-out mode where an individual forms a group by inviting other participants to a session; and dial-in mode, where individuals join an existing Chat channel.

The Dial-Out participant method includes:

1. One to One session
2. Ad-hoc group session
3. Pre-arranged group session
4. 1 to Many to 1 group session.

The Dial-In method includes:

1. Open Chat session
2. Restricted Chat session.

OMA PoC groups and session types are given in Table 1.2.

When a dial-out PoC session type is used, the PoC service sends an invitation request to each intended session participant (each member of the group). In order to complete session establishment, call acceptance by at least one intended callee is required. There are two ways to implement this:

1. The end user being manually alerted before accepting the session invite;
2. The end users' device automatically accepts the session invite.

Alerting is the common method for multiparty conference calls, where the invited party sees who is inviting them to the session. Manual alerting corresponds with an On Demand session setup, as the recipients may or may not accept the session invite. It also corroborates with the notion of the PoC Session Owner, as the session can be dropped when the initiator leaves the group call. The automatic session accept assumes that the recipient does not need to see who is calling and is suitable for professional PoC users who need a walkie-talkie type call setup. In general, all the participants are known, and the calling user is thus placed into the Access Control *Accept List* to enable fast call setup. Automatically accepting the call combined with the Pre-established Session feature, optimizes the call setup time.

The Dial-In PoC session types do not require alerting, as PoC users themselves decide to join or leave Chat channels. The PoC user has the option to be simultaneously attached to many channels and receive PoC speech whenever a speaker from one of the sessions sends a talk burst (*Paivi says 'hi there' from channel 3*). Further, as there is no alerting, a PoC user

Table 1.2 Push-to-Talk over Cellular group and session types

Concept	OMA Definitions
PoC Group – A predefined set of PoC users together with its attributes. A PoC group is identified by a Session Initiation Protocol Uniform Resource Identifier, or SIP URI (PoC Group Identity for PoC Pre-arranged and Chat PoC Groups). PoC users use PoC Groups to establish PoC Group sessions	
PoC Group Identity	A SIP URI identifying a Pre-arranged PoC group or a Chat PoC group. A PoC Group Identity is used by the PoC client to establish PoC group sessions with the Pre-arranged PoC groups and Chat PoC groups
Pre-arranged PoC Group	A Pre-arranged PoC group is a persistent PoC group that has an associated set of PoC users/and PoC groups. The establishment of a PoC session with a Pre-arranged PoC group results in all members being invited
Chat PoC Group	A persistent PoC group in which each member individually joins the PoC session. The establishment of a PoC session with that PoC group does not result in other members of the Chat group being invited
Restricted Group	A group that can be joined only by a PoC user that is a member of the group. A Restricted group has a Group List
PoC Group session – is a Pre-arranged PoC group, Ad-hoc PoC group or Chat PoC group session. A Session Type is a SIP URI- parameters are used to convey the type of SIP URI, and may take on one of the following values: 1-1, Ad-hoc, Chat Pre-arranged or Chat.	
1-1 PoC Session	A feature enabling a PoC user to establish a PoC session with another PoC user
Ad-hoc PoC Group Session	A PoC group session established by a PoC user to form an Ad-hoc PoC group
Pre-arranged PoC Group Session	A PoC session established by a PoC user with a Pre-arranged PoC group
Chat PoC Group Session	A Chat PoC group session is a PoC session established with a Chat PoC group

may automatically connect to channels that are active when he/she goes online. To differentiate between the channels, there is a Primary PoC Session (*the favourite channel*) and Secondary PoC Session (*a list of channels the user wishes to interact in*).

As PoC sessions are an online service, both the invite (dial-out) and join in (dial-in) group attachments benefit from the Presence service. The Presence service shows who is online and is available for communication. Once seeing Mary and Alice are online, John may initiate an Ad-hoc session with both.

OMA PoC session concepts are presented in Table 1.3.

1.3.1 One to One Session

One to One sessions are PoC calls set up by an originator to a single user instead of a group. John creates a One to One session during a robotics conference by a selecting participant,

Table 1.3 Push-to-Talk over Cellular Session definitions

Concept	OMA Definition
PoC Session Owner	In the case of 1-1 PoC session and Ad-hoc group session, the PoC Session Owner is the initiator of the PoC session. In the case of a Chat PoC group and a Pre-arranged PoC session, the PoC Session Owner is the creator of the PoC group
PoC Session Identity SIP URI	Identitifies the PoC session and can be used for routing initial SIP requests. It is received by the PoC client during the PoC Session establishment in the contact header and/or TBCP Connect message in the case of a Pre-established Session
Pre-established Session	A Pre-established Session between the PoC client and the PoC server. The PoC client established the Pre-established Session prior to making requests for PoC sessions to other PoC users. To establish a PoC Session based on a SIP request from a PoC user, the PoC server conferences other PoC servers/PoC users to the Pre-established Session to create an end-to-end connection. *This is relevant to an automatically accept answer mode.*
On Demand Session	An On Demand Session is a PoC session setup mechanism in which all media parameters are negotiated at PoC session establishment. *This is relevant to a manual alert answer mode*
Simultaneous Session	This is where a PoC user is a Participant in more than one PoC session, simultaneously using the same PoC client. When a PoC user is participating in Simultaneous PoC Session, the PoC server performing the Participating PoC Function will apply media filtering on continuous media types (voice, video) between PoC sessions providing the same media.
Primary PoC Session	This is the PoC session that the PoC user selects in preference to other PoC sessions. When the user has simultaneous PoC sessions, the Primary PoC Session has priority over Secondary PoC Sessions
Secondary PoC Session	This is a PoC session for which the user receives media when there is no media present on the Primary PoC Session

'Mary' from his contact list and presses the PTT key (Figure 1.2). If Mary is not available, John can send her an Instant Personal Alert (*a call-back message*). John's Instant Personal Alert (IPA) will be stored in Mary's 'Call Back Box'. If Mary does not want to be disturbed by John, she can choose to block his call by selecting *Incoming Session Barring (ISB)* or *Incoming Instant Personal Alert Barring*.

The One to One call session starts when the call is initiated and talking is mediated by floor control. When John presses the PTT key to speak, a floor request is sent to the PoC service. The request is acknowledged by the PoC service with a floor grant and John starts to speak – '*Hello, Mary, how are you doing?*' When John ends the transmission, his mobile device sends a floor release indication to release the floor. If Mary tries to interrupt John when he has the floor, the PoC service will deny her request, indicating that she needs to wait until John stops talking. When John releases the floor, (stops pressing the PTT key) the floor is

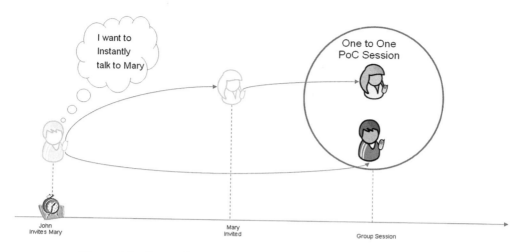

Figure 1.2 One to One Session – John invites Mary for an low cost call

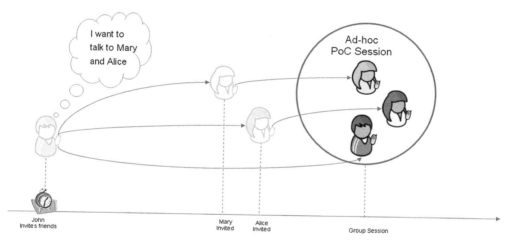

Figure 1.3 Ad-hoc Group Session – John invites Mary & Alice to agree where to meet and when

idle until Mary presses her PTT key and replies *'I'm fine, John, I have a favour to ask'*. The session ends after a period of inactivity or when explicitly closed by John or Mary.

1.3.2 Ad-hoc Group Session

Ad-hoc sessions are temporary group calls set up by an originator to establish a group session with no requirement of an existing permanent group. John creates an Ad-hoc group during the robotics conference by selecting participants one by one from his contact list and presses the PTT key (Figure 1.3). The mobile device sends the list to the network during the invitation; however the network does not store the list. As John and other online PoC users are at

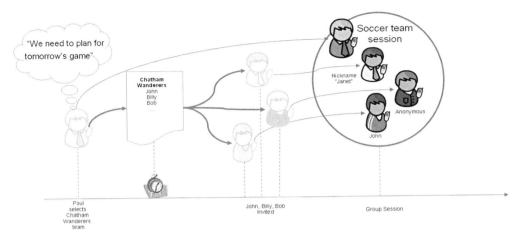

Figure 1.4   Pre-arranged Group Session – Paul invites the 'Chatham Wanderers' for team talk

a conference with people they hardly know, they have set their mobile devices to manual alert – so they can see who is calling them before accepting the PoC call. They also use Incoming Session Barring (ISB) during lunch breaks and set the ACL Reject Lists to stop some participants from constantly alerting them.

During the PoC session, John can periodically check the identity of the other members in the group and request indication of group participant changes. For example, during the Ad-hoc PoC session with Mary and Alice, John is sent a notification when Alice leaves the session.

1.3.3 Pre-arranged Group Session

A Pre-arranged group is a group where access is limited to predefined list of participants. This list is created by the group owner, and stored by the service provider by means of a specific XML document. For example, Paul has set up a Pre-arranged sports group called 'Chatham Wanderers' (Figure 1.4) and listed all of the team members using his mobile device. This PoC XML group document is stored in the Wirelessfuture.com network. When Paul selects the 'Chatham Wanderers' name on his mobile device, the Wirelessfuture.com PoC service will invite the all of the online team members to participate in a group session. As soon as Paul presses the PTT key, the invited team members will receive an indication of the incoming session, including the identifications of the 'Chatham Wanderers' group and origi-nator 'Paul'. The invited team members join the group session either with an automatic or manual reply. Paul will receive a notification of the invited members who have accepted the invitation. If at least one of the invited team participants accepts the invitation, the group session enters the communication phase and any of the participants that are present can start communicating.

As Paul is the group owner, he can invite absent group members to join the session while it is in progress. However, the Chatham Wanderers team members may join, leave, and rejoin the existing session without affecting the group discussion. Termination of the session occurs when Paul, the Session Owner, leaves the session or if the number of remaining participants in the session drops to one. Session Termination may also occur if there is an inactivity timeout.

1.3.4 Open Chat Group Session

An Open Chat group is a group to which anyone who knows the group identification can join. John has joined an Open Chat group called sip:*pocquiz@games.wirelessfuture.com*, he found the group identification from a gamers' web site named *finders*. The purpose of the gaming group is to answer questions related to the location of an object using PoC. The participant who answers the most questions within a defined time wins a cash prize. The pocquiz Open Chat group is advertised once a week to PoC users who have the free time to join the game. The service provider, *Wirelessfuture.com* charges a special rate for PoC gamers.

1.3.5 Restricted Chat Group Session

A Restricted Chat group is a group where access is limited to predefined participants. John is a part-time employee of an organization called *Enterprise Forwarding*. *Enterprise Forwarding* delivers freight throughout Europe. A local Finnish IT management user has assigned John the rights to talk and listen to members of '*Italian_deliveries*'. He has also given him the rights to create Ad-hoc professional sub groups within the '*northern sector*' delivery region.

John joins the *Italian_deliveries* group channel (Figure 1.5) by selecting the name from the contact list in his mobile device when he starts work and stays connected throughout

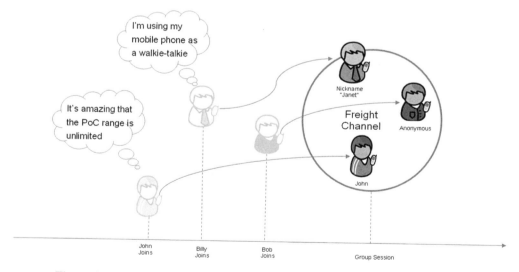

Figure 1.5   Restricted Group Chat – John, Billy and Bob join the Drivers' Session

until the workday finishes. He sets his mobile device to automatic answer so he hears the drivers' talk bursts as soon as they start giving information of their whereabouts. He also sees the drivers' talking party identification. He asks the drivers questions by pressing the PTT key. For special weekend deliveries, John creates his own Restricted Chat groups and advertises the group ID to selected drivers. At the end of the workday, John leaves the '*Italian_deliveries*' group talk session without affecting the group session.

1.4 Multimedia Group Communication Implementation

The implementation of multimedia group communication – OMA PoC, XDMS, and Presence with SIMPLE enablers – lies within the domain of computer networking and Internet Protocols (IP). These enablers reuse Internet Protocols – Session Initiation Protocol (SIP), Real-time Protocol (RTP), Real-time Control Protocol (RTCP), and Hypertext (HTTP) – which makes them relatively easy for vendors to develop. Vendors either develop PoC, XDMS and Presence clients, primarily for mobile phones, or PoC, XDMS, Presence application servers.

The interaction between the end user's actions and the application of Internet Protocols are shown in Figure 1.6. SIP is used for call setup set and tear down (signalling), RTCP is used to control the media flow and the right to speak, and RTP is used for PoC Speech transport:

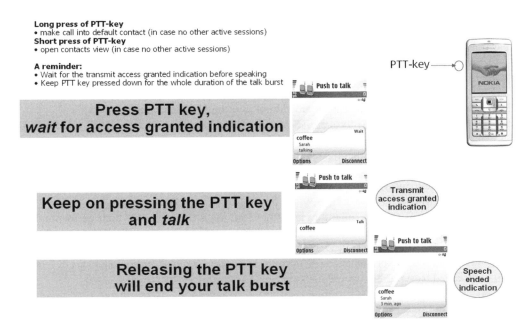

Figure 1.6   Steps on using a prototype PoC Application (2005)

- 'Long Press of PTT-Key' to make a call involves PoC Signalling by SIP to the SIP/IP core, PoC server and to other participants;
- 'Wait for access granted indication' involves receiving RTCP and Talk Burst Control Protocol (TBCP) messages from the PoC server;
- 'Keep on pressing the PTT key and talk' involves RTP transmission to the PoC server and on to the other participants;
- 'Releasing the PTT key will end your talk burst' involves sending RTCP and TBCP release message to the PoC server. The floor becomes idle and other participants have the opportunity to talk. After a period of inactivity, session is released by SIP signalling.

In the same way that Internet resources are assigned by Uniform Resource Identifiers (URI), the PoC service, XDMS, and Presence services use URIs to address PoC users, PoC groups, PoC sessions, The PoC service, XML documents, and Presence entities (Table 1.4).

The next sections will elaborate more on PoC SIP signalling, PoC Speech, XDMS and Presence signalling.

1.4.1 PoC Signalling

PoC Signalling is based on the Session Initiation Protocol (SIP). SIP is an application-layer control protocol for creating, modifying and terminating sessions with one or more participants. PoC SIP clients use TCP or UDP (typically on port 5060) to connect to SIP servers (IMS & PoC servers) and other SIP endpoints (PoC clients).

PoC Signalling uses a range of SIP methods and messages to mimic the sequence of a circuit switched telephone call, i.e. call setup, call acknowledgements and call tear down.

In general, all PoC sessions follow the following sequence. The SIP methods are highlighted in capitals [2]:

1. A PoC client (SIP User Agent) is embedded into the User Equipment;
2. A PoC client registers to the service providers PoC service via the SIP/IP Core (IMS) using the SIP REGISTER method;
   a. SIP/IP Core (IMS) authenticates the PoC user and routes the SIP messages to the PoC server;
3. The PoC client informs the PoC server about the client's capabilities and the PoC user preferred settings to PoC service by sending a SIP PUBLISH message;
4. Upon user action, the PoC client invites other PoC users (identified by their corresponding PoC identities) to a PoC session or joins a PoC Channel using an SIP INVITE;
   a. SIP INVITE with Session Description Protocol (SDP) informs the PoC server about the client media parameters to be used in the session, such as audio codec or UDP ports allocated to send/receive media;
5. During a PoC session, PoC users are informed of other PoC users' changes by means of SIP NOTIFY messages;
   a. The PoC server informs group Participants who is in attendance;
6. PoC users leave the PoC session sending a SIP BYE.

Table 1.4　Push-to-Talk over Cellular Identities

Concept	OMA Definition
SIP URI	From RF3261: A SIP or SIPS URI identifies a communications resource [1] and follows the guidelines in RF2396. PoC uses SIP URI's to identify PoC clients, PoC servers, and PoC sessions, resource lists that point to URI lists etc . . . '*sip:john@ wirelessfuture.com*' is a SIP URI
TEL URI	A TEL URI identifies a resource identified by a telephone number. '*tel:+34-600123456*' is a TEL URI
PoC User Identity	This is the identity of PoC subscribers. A PoC user identity may take the form of either a SIP URI or a TEL URI. A given subscriber may have assigned one of more PoC identities, of which at least one must be of the form of a SIP URI. The PoC user identity is used to address remote contacts, or to show the identity of the calling user
Session Type	A SIP URI- parameters are used to convey the type of SIP URI, and may take on one of the following values: 1-1, Ad-hoc, Pre-arranged or Chat
PoC Session Identity SIP URI	Identitifies the PoC session and can be used for routing initial SIP requests. It is received by the PoC client during the PoC session establishment in the contact header and/or TBCP Connect message in the case of using a Pre-established Session. For example: '*sip:session12345@poc-server.wiressfuture. com;session=1-1*'
Conference-Factory-URI	A provisioned SIP URI that identifies the PoC service in the Home PoC Network, typically used for setting up an Ad-hoc PoC group or 1-1 PoC session. For example: '*sip:ad-hoc@ wirelessfuture.com*'
PoC Group Identity	A SIP URI identifying a Pre-arranged PoC group or a Chat PoC group. A PoC Group Identity is used by the PoC client to establish PoC Group sessions to the Pre-arranged PoC groups and Chat PoC groups. For example: '*sip:john@wirelessfuture. com;poc-group=mypocgroup*'
Exploder URI	An address of a SIP URI-list service. A URI-list service is a specialized application service that receives a SIP request with a list of Uri's and generates a similar SIP request to each of the Uri's on the list. The SIP URI-list service includes a copy of the body of the original SIP request in the generated SIP requests
Anonymous PoC Address	A PoC address that may be used by PoC clients (or servers) to hide the identity of the end user, when privacy has been requested. An anonymous address may be a valid SIP URI, where the username and host part do not contain any valid information about the final user it represents

There are other SIP methods sent to interact with the PoC service, such as CANCEL, PUBLISH, MESSAGE, REFER ... The dialog also includes a variety of SIP messages, such as Information (100 Trying, 180 Ringing ...), Success (200 OK), Client Errors (401 Unauthorized, 403 Forbidden, 408 Request Time Out), Server Errors (501 Not Implemented) ...

Outside the session, PoC users can send SIP messages to inform other users of the existence of a newly created group or to ask a specific user to call back [1][2]:

- Group Advertisement
- Instant Personal Alert.

Figure 1.7 shows the SIP messages used to control PoC Communications. The PoC client is a SIP User Agent and transmits SIP messages to the SIP/IP Core (IMS in this example). The IMS transmits and received SIP messages to the PoC service. The IMS transmits the SIP messages to the invited parties.

There are two important phases in PoC Signalling. The first is when the PoC user's client informs the PoC service about its capabilities and the PoC user's defined settings – such as *'do not disturb' with Instant Personal Alerts* and today my nickname is *'Susan'*. The second important phase concerns media negotiations between PoC clients during session setup. Codec negotiation between originating PoC client, PoC server and terminating PoC client is an area where interoperability issues may arise.

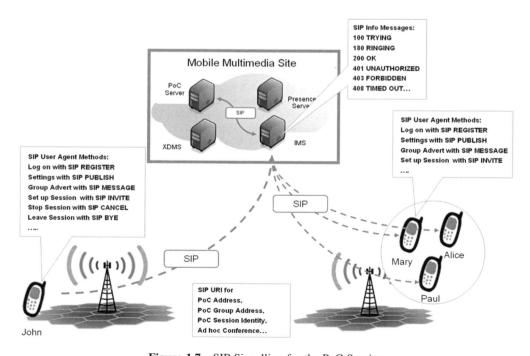

Figure 1.7   SIP Signalling for the PoC Service

1.4.1.1 PoC Client settings

Following SIP REGISTRATION to the PoC service, the PoC client PUBLISHES the capability of the PoC client and the willingness of the PoC user to convey with related PoC client and PoC server functionalities. The PoC client settings are sent as an XML payload and include (Table 1.5).

- Answer Mode – The current Answer Mode setting preference of the PoC user
- Incoming Session Barring – activated/not activated
- Incoming Personal Alert Barring – activated/not activated
- Simultaneous PoC Session Support.

1.4.1.2 CODEC Negotiation

During session setup, SIP/SDP-based information exchanged between the PoC server and the PoC client specifies the Media Parameters for a PoC session being established or that already exists (in the latter, Media Parameters may be re-negotiated in certain cases).

Table 1.5 Push-to-Talk over Cellular Client Service Settings

Concept	OMA Definition
PoC Client	A PoC functional entity that resides on the PoC User Equipment and supports the PoC service
PoC Address	An address that identifies a PoC user and is used by the PoC client to request communication with other PoC users
PoC Service Settings	
Automatic Answer Mode	A PoC client mode of operation in which the PoC client accepts incoming PoC session establishment requests without manual intervention from the PoC user; media is immediately played when received
Manual Answer Mode	A mode of operation in which the PoC client requires the PoC user to manually accept the incoming PoC session before the PoC session is established
Incoming Session Barring	Conveys the PoC user's desire for the PoC service to block all incoming PoC session requests
Incoming Instant Personal Alert Barring	Conveys the PoC user's desire for the PoC service to block all incoming Instant Personal Alerts
Simultaneous PoC Session	Conveys that the PoC client is able and PoC user is willing to use Simultaneous PoC sessions
PoC User Defined Settings	
Nickname	A user-friendly display name that might be associated with a PoC user or a PoC group. This can be the Nickname that is displayed to other users in the channel
Anonymous PoC Address.	A PoC Address identifies a PoC user who has requested privacy
Instant Personal Alert	A feature in which a PoC user sends a SIP based instant message to a PoC user requesting a 1-1 PoC session. *Call back request tone – define the ringing tone for call back requests*

The AMR codec modes are negotiated at the start of the session. They are controlled in the server where the session is originated [2].

1. A PoC client sends a list of the codec modes that it supports to the PoC server;
2. The PoC server compares the codec modes sent by the client to the modes listed in the PoC server parameters and makes a common subset. This subset is then returned to the client;
3. The client implementation determines which one of the available codec modes is chosen.

Each PoC client performs codec mode negotiation with the PoC server individually. Even if an identical subset is sent by the PoC server to different terminals, the terminals might choose to use different codec modes. The negotiated AMR codec modes are not changed during the session.

1.4.2 PoC Speech

PoC Speech is transmitted in Real-time Transport Protocol (RTP) packets. RTP is an IP-based protocol providing support for the transport of real-time data such as audio or video streams. RTP is designed to work in conjunction with the auxiliary control protocol RTCP to get feedback on the quality of data transmission and information about participants in the ongoing session. In PoC, RTCP is also used to carry Talk Burst Control Protocol (TBCP) messages that are used to arbitrate requests from PoC clients for the right to send media – *in other words, floor control* [3].

RTP is typically run on top of UDP to make use of its multiplexing and checksum functions. To provide timely delivery of PoC Speech, RTP provides time stamping, sequence numbering, and other mechanisms to take care of the timing issues. The receiving PoC client uses the time stamp to reconstruct the original timing in order to play out the payload data (PoC Speech) in correct rate. Sequence numbers are used to place the incoming payload data packets in the correct order. From the payload type identifier (e.g. what CODEC was used to encode transported data), the receiving application knows how to interpret and play out the payload data. The RTP source identification allows the receiving application to know where (e.g. which PoC user) the data is coming from.

The flow of PoC speech from a PoC client is known as a Talk Burst.

The general flow for exchanging talk bursts between participants is as follows:

1. SIP establishment phase
   a. SIP INVITE with Session Description Protocol with media parameters;
   b. UDP port numbers are identified as endpoint of the media packets (unique port numbers for each PoC session) and connection is set up;
2. Inviting PoC user is granted first talk burst
   a. Inviting PoC user receives TBCP Granted
   b. Invited PoC users receive TBCP Taken;
3. Inviting PoC user talks – transmits a talk burst
   a. Real-time data (PoC Speech) sent in RTP payload;
4. Invited PoC users receives talk burst;

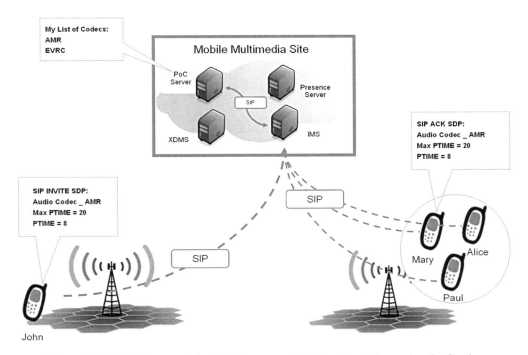

Figure 1.8 PoC Clients and the PoC Server negotiate CODEC to be used in the Session

5. Talking user stops talking and releases the button
   a. The rest of the users receive a notification that the floor is idle now, and the process reinitiates again at 2, as soon as any user requests the right to send a talk burst;
6. After a period of inactivity the RTP session is closed
   a. TBCP Idle.

Figure 1.9 shows the flow of RTP Messages used to route PoC Speech from the originator to the terminating parties. The PoC client transmits RTP messages directly to the PoC server. The PoC server transmits the PoC Speech directly to the terminating parties. RTCP is used for statistics. TBCP is used for floor control and gives indications when the floor is free. Table 1.6 lists the various dialogs a user receives during a group session.

1.4.3 XML Document Management Signalling

XML Document Management uses HTTP to make remote connections. HTTP is a request/response protocol between clients and servers. The high level concepts of OMA XDMS are shown in Figure 1.10 and Table 1.7. The general uses for XDMS are described below:

1. XCAP functionality (HTTP client and XML document manager) is embedded into the User Equipment;

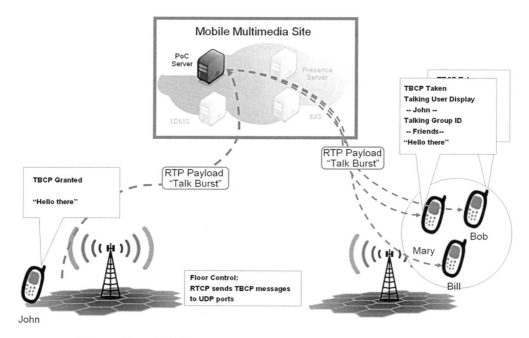

Figure 1.9   Media flow between group Participants is transported by RTP

2. End users create lists and upload XML documents (PoC Group Document or Access Control List, Presence List) using a HTTP client to connect to the XDMS server (a HTTP server);
   - All users are authenticated by the XDMS Aggregation Proxy;
3. XML documents (identified by HTTP URI) are routed to the relevant service, e.g. the PoC server XDM component;
4. End users download XML Documents for modification and deletion;
5. The same XML Document (URI list) can be shared with other services.

1.4.4 Presence Signalling

Presence Signalling between client and server is based on SIP/SIMPLE (Figure 1.11, Table 1.8). The high level concepts of OMA Presence are described below:

1. Presence client (SIP User Agent) watcher functionality is embedded into the User Equipment. This feature lets a client receive notifications about remote contacts, such as their availability for PoC communication.
2. Presence user registers to the Presence service via the SIP/IP Core (IMS) using SIP REGISTER.
3. Presence users (or Presentities) announce their availability using SIP PUBLISH. SIP PUBLISH contains an XML document that expresses Presentity's status. Eventually,

Table 1.6 Push-to-Talk over Cellular Client Displays and Indicators

Indicator	Usage
Talking User ID display	In a one-to-one call, identification (SIP URI, and/or mnemonic) of the individual user, whose voice is currently received, is displayed. In a group call, user identity is displayed together with the group ID
Talking Group ID display	In group calls, the identification of the group (URI), where the voice is currently being received from, is displayed
Instant Personal Alert	Indicator informs the user about the received Instant Personal Alert in a mobile device
Group Advertisement	Indicator informs the user about the received Group Advertisement in a mobile device
Incoming Call Notification	Incoming call notification indicator informs the user about an incoming notification in a mobile device
Notifying	A notifying indicator informs the user, who is trying to make a one-to-one call, that the call attempt has been turned into a notification towards the target user
Call Not Allowed	Call not allowed indicator informs the user, who is trying to make a one-to-one call that the call attempt to the target user is not allowed. The indicator is used in situations where the network does not authorise the call. The recipient is not notified about the call attempt
Ready To Talk	Ready to talk indicator is an indication appearing after pushing the Push-to-Talk key and telling that the user can start talking. It means that there is a high probability for a successful connection
End of Talk Burst	End of talk burst indicator is an indication to the receiving user that the other user has released the Push-to-Talk key, and the others can take turns. The originating user also gets the indication
Time Exceeded	Time exceeded indicator is an indication signifying that the maximum talk burst time has been exceeded and the talk burst is interrupted. To continue with a new talk burst, the user needs to release and press the PTT key again
Transmission Failed	Transmission failed is an indication signifying that the transmission attempt has failed. The reason for the failure can be, for example, that the downlink connection to the recipient could not be established in a one-to-one call or talk burst collision
Incoming Call	A new incoming call is preceded by an indicator. A new call means a voice transmission after a period of inactivity in the communication. It can also mean a changed communication party in a call

information contained in the SIP PUBLISH message is transferred to all watchers interested in receiving Presence information about this user.

4. Presence users (Watchers) discover availability of other Presence users (online/offline status) using the SIP SUBSCRIBE message. SUBSCRIBE places a Presence *subscription*. While the subscription is active, the Presence service delivers notifications (SIP NOTIFY) to all subscribed watchers, to inform about changes in the status of each Presentity.

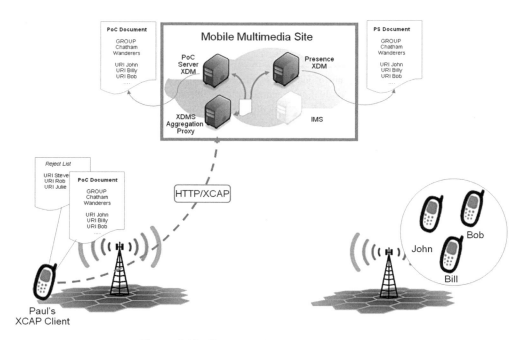

Figure 1.10   Document Management via XDMS

Table 1.7   XML Document Management Concepts

Concept	OMA Definition
URI List	A collection of URI's put together for convenience
Group Management	The action of creation, modification or deletion of XML documents that define groups. Groups can in turn be used by SIP services such as PoC or Presence. XML documents are stored in a XDM server such as the PoC XDMS or the Shared XDMS.
Group Usage List	A list of group names or service URI's that are known by the XDM client. This can be thought as an XML document that stores the list of all PoC groups known to a XDMC. In the future, it may contain group identities related to other services (e.g. Messaging)
XCAP Client	An HTTP client that understands how to follow the naming and validation constraints defined in the XDML specifications
XML Document Management Server	An HTTP server that understands how to follow the naming and validation constraints defined in the XP specification
Shared Group XDMS	The Shared XDMS is an XCAP server that manages XML documents (e. g. URI Lists), which can be shared among several enablers.
Application Unique ID (AUID)	A unique identifier that differentiates XCAP resources accessed by one application from XCAP resources accessed by another.
Document URI	The HTTP URI containing the XCAP root and document selector, resulting in the selection of a specific document
Node URI	The HTTP URI containing the XCAP root, document selector, node selector separator and node selector resulting in the selection of a specific XML node

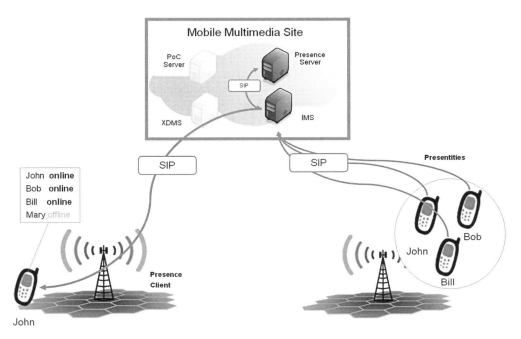

Figure 1.11   Presence Service using SIP signalling

Table 1.8   Presence with SIMPLE concepts

Concept	OMA Explanation
Presentity	A logical entity that has Presence information associated with it. This Presence information may be composed from a multitude of Presence Sources. A Presentity is most commonly a reference for a person, although it may represent a role such as 'help desk' or a resource such as 'conference room #27' The Presentity is identified by a SIP URI (as defined in R3261), and may additionally be identified by a pres URI (RF3859).
Presence User Agent	A terminal or network located elements that collects and sends user related Presence information to a Presence server on behalf of a Principal Source
Watcher	Any uniquely identifiable entity that requests Presence information about a Presentity from the Presence service.
Subscriber	A form of watcher that has asked the Presence service about a subscriber's request to be notified of changes in the Presence information of one or more Presentities
Fetcher	A form of watcher that has asked the Presence sservice for the Presence information of one or more Presentities, but is not requesting a notification from the Presence service of (future) changes in a Presentity's Presence information

Table 1.8   *(continued)*

Concept	OMA Explanation
Subscribed-watcher	A type of watcher, which requests notification from the Presence service of changes in a Presentity's Presence information, resulting in a watcher-subscription, as they occur in the future
Poller	A fetcher that requests Presence information on a regular basis
Presence Infrastructure	
Presence Server	A logical entity that receives Presence information from a multitude of Presence Sources pertaining to the Presentities it serves and makes this information available to watchers according to the rules associated with those Presentities
Resource List Server (RLS)	A functional entity that accepts and manages subscriptions to presence lists, which enables a Watcher application to subscribe to the Presence information of multiple Presentities using a single subscription transaction
Presence Network Agent (PNA)	Network located element that collects and send network related Presence information on behalf of the Presentity to a Presence server

1.5 Summary and Conclusions

This chapter has given the reader an overview of the main concepts associated with multi-media group communication for the mobile domain. Essentially, Group Communication consists of two components: list management and execution of the group function. List Management concerns administrating the group roles – defining ownership and membership. Execution of a group function concerns the invocation of a PoC session that occurs either in dial-out mode or in dial-in mode. Interaction within a group session is conditioned by floor control in which only one participant can talk at a time.

The implementation of OMA PoC, XDM and Presence with SIMPLE is more akin to mobile computer networking and Internet Protocols than with circuit switched group telephony. OMA PoC is VoIP technology that emulates a walkie-talkie experience and is based on SIP signalling and RTP transport. XDMS uses HTTP to transport XML documents. Presence with SIMPLE uses to SIP signalling to express the willingness of online subscribers to communicate.

The rest of this book will expand on the above principles.

1.6 References

[1] OMA Push-to-Talk over Cellular (PoCv1.0.2): 'Push-to-Talk over Cellular Requirements Document'; June 2006.
[2] OMA Push-to-Talk over Cellular (PoCv1.0.2): 'Push-to-Talk over Cellular Control Plane'; September 2007.
[3] OMA Push-to-Talk over Cellular (PoCv1.0.2): 'Push-to-Talk over Cellular User Plane'; September 2007.
[4] OMA XML Document Management (XDMv1.0.1): 'XML Document Management Requirements Document'; June 2006.
[5] OMA Presence SIMPLE (Presence v1.0.1): 'Presence SIMPLE Requirements Document'; July 2006.

2

OMA Push-to-Talk Architecture

2.1 Introduction

Push-to-Talk over Cellular (PoC), as an IP-based service, can be considered the first Voice-over-IP (VoIP) technology delivered in a broad manner over cellular networks worldwide. An important difference between PoC and regular voice communications is that it is half-duplex based, converting it into a *walkie-talkie* style of service, with the advantage over regular walkie-talkie that it enjoys ubiquitous coverage provided by the cellular network infrastructure.

In order to setup PoC communications, *Session Initiation Protocol* (SIP) signalling is used between the user equipment (UE) and the network. *Real-time Transport Protocol* (RTP) delivers digitized voice to all participants in a session. An additional mechanism – the *Talk Burst Control Protocol* (TBCP)– is used to determine which user has the right to speak to the rest of participants in a PoC session at any given time. SIP and RTP are well-known protocols in the multimedia architecture defined by IETF, while TBCP is a specific extension defined by OMA to support the PoC service in particular.

The OMA PoC enabler defines a set of network functions and their interaction in an open environment. In particular, usage of IETF standards-based technologies ensures that handsets and network equipment supplied by different vendors can interoperate seamlessly, thus avoiding fragmentation and benefiting the end user, who is able to communicate with other peers, as long as they use another OMA compliant client implementation.

Although other non-standard PoC pre-standard implementations have been commercially deployed, this chapter will focus on describing the architecture of the standard OMA PoC system. It is expected that most legacy PoC deployments will progressively converge towards the OMA PoCv1 technology as the common baseline for true interoperability across devices and networks.

Figure 2.1 shows a high-level overview of the logical functions required to deliver a PoC service [1]. The two main entities supporting a PoC service are a PoC-enabled handset and the PoC service network entity: the PoC server.

The PoC service requires some common functionality such as service discovery, user registration, authentication, service authorization and security features. These are requirements

Multimedia Group Communication. Andrew Rebeiro-Hargrave and David Viamonte Solé
© 2008 John Wiley & Sons, Ltd.

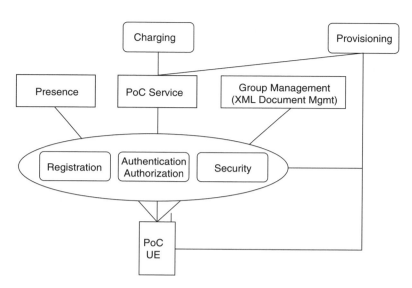

Figure 2.1   OMA PoC Logical Architecture

shared with other services as well, so typically an external underlying common platform provides these services to PoC [1]. OMA defines such platform as a generic SIP/IP Core infrastructure, which – in 3GPP / 3GPP2 deployments – maps to the *IP Multimedia Subsystem* (IMS).

In order to support group communications, PoC requires services from the Group Management entity, defined by OMA as the *XML Document Management* (XDM) enabler. The XDM enabler manages XML documents stored in the network. As such, XML documents can define PoC groups or user-defined policies, such as *blacklisted* addresses. The only difference between an XML document defining a PoC group and an XML document containing a service policy may be the actual format (XML schema) and content of the information. Hence, XDM specifies a generic mechanism to manage XML documents. These documents, in turn, accomplish different types of purposes, which are generally dependant on the final application (e.g. PoC, Presence, Messaging) which makes use of them.

The third application function that we can observe in Figure 2.1 is the *Presence* feature. The Presence service provides PoC users with the ability of determining whether their intended contacts are available for communication *prior* to establishing a PoC session. Presence is useful, therefore, to let users decide the most appropriate – and less intrusive – means of communication at any given time. Internet users are becoming increasingly familiar with Presence solutions. Effectively, the Presence user experience is already available in Internet communication tools that support services such as Instant Messaging or voice communications. OMA has defined a Presence service based on the IETF SIMPLE[1] framework.

---

[1] SIMPLE stands for *SIP for Instant Messaging and Presence Leveraging Extensions*, and is the IETF Working Group in charge of defining extensions to the Session Initiation Protocol in order to support Presence and Messaging features.

OMA Presence is an optional enabler: from the technical point of view, OMA Presence is not strictly required to deliver PoC services. However, given its benefit in terms of enabling a smooth and non-intrusive user experience, it is expected that most OMA PoC deployments will enjoy Presence functionalities as well.

Users interact with the PoC service with a PoC-enabled handset (the PoC User Equipment). Typically, the PoC UE will also implement XDM features to interact with the XDM enabler and let the user manage groups and policies. In such a case, the PoC UE incorporates XDM client functions (XDMC). Finally, when Presence features are used to enhance the PoC user experience, the PoC UE is able to retrieve and display Presence information from other users (to let the PoC user know the availability and status of their PoC contacts).

In general, the service provider must support at least two additional functionalities to deploy a commercial PoC service: service provisioning and charging. Service provisioning must take place in order to ensure proper configuration of all elements in the end-to-end path, including the PoC server, the PoC client, the SIP/IP Core and the access network. PoC charging is based on the information delivered by the PoC server towards an external charging entity to enable end-user charging and billing.

Before proceeding with the detailed description of all elements involved in the PoC architecture, and in order to ease reading, it is worth presenting the four basic protocols required to support the PoC service:

- The Session Initiation Protocol (SIP) [2] is the main signalling protocol used between PoC clients, the SIP/IP core and the PoC service entity. As such, it supports basic operations such as user registration, one-to-one and group PoC session initiation or sending of *call-back request* messages to other parties.
- The Real-time Transport Protocol (RTP) [3] supports transport and delivery of encoded and packetized voice samples over an IP infrastructure. The Real-time Control Protocol (RTCP) is generally used together with RTP to provide information about the quality of the received stream, such as lost packets statistics, average delay and jitter, . . .
- The Talk Burst Control Protocol (TBCP) [4]. Given the half-duplex nature of PoC, TBCP is the OMA-defined protocol used to request the right to speak during a PoC conference. The PoC server is the one arbitrating user requests and deciding who has 'the floor' (i.e. the right to speak) at a given time. TBCP is an extension to RTCP [3] that makes use of RTCP *Application* packets (APP) to arbitrate floor control requests.
- The XML Capability Access Protocol (XCAP) [5] lets users create, modify and delete XML documents stored in a XDM server. In particular, users may use XCAP to create permanent PoC groups, which are then used by the PoC service to establish group communications.

In the following sections we will provide a more detailed description of all the elements involved in the PoC service architecture.

2.2 Architectural Considerations

Before presenting the technical structure of the OMA PoC service, it is worth describing the paradigms that were taken into consideration when designing the PoC service architecture, and that eventually led to the final architecture defined in [1]. These paradigms are briefly described below.

- OMA PoC is a *SIP-based service*. SIP [2] is one of the areas of greatest interest and growth in both the Internet and mobile communities, where future multimedia services and technologies are based on SIP. PoC, as a new communication paradigm for cellular users is, in fact, the entry point for new SIP-based services to come and, as such, it has been clear since the early days of OMA PoC standardization that PoC would certainly be a SIP-based service.
- OMA PoC relies on IETF specifications to ensure alignment with commonly accepted Internet technologies. When possible, OMA PoC does not try to *reinvent the wheel*, unless a required functionality cannot be delivered by existing IETF work within intended timeframes. This is true not only for PoC, but also for other OMA enablers, as we will see throughout this book.
- OMA PoC is an *open standard*. OMA produces open specifications, which are available for public review and implementation.
- OMA has *interoperability* and *implementability* among its most important goals. OMA architecture is based on a set of well-defined logical functions connected by open interfaces, based on standard technologies such as SIP, RTP or XCAP. This openness ensures interoperability in a multi-vendor environment, and the capability of connecting the PoC services with other OMA-defined enablers, such as Presence or XDM.
- OMA PoC should be a scalable and efficient service. When feasible, OMA tries to ensure that design considerations scale up to full commercial deployments supporting millions of users.
- Finally, an imporant goal for OMA PoC (and Presence and XDM) was to produce a specification that could be implemented on top of existing and future IP-based cellular networks. This requirement means that OMA enablers can be deployed over any underlying network technology, if it supports some basic functionality such as SIP routing and IP connectivity. Additionally, OMA takes special care that OMA enablers interoperate in a smooth and seamless way with IP multimedia technologies, such as IMS, as defined by the main cellular standardization bodies, namely: 3GPP and 3GPP2 [19].

In general, OMA and 3GPP have defined a general split in the scope of each organization's activities, with OMA focusing on the service and service enabler layer, and 3GPP focusing on cellular and core network technologies. Thus, OMA has become the natural environment for the standardization of future mobile multimedia services.

2.3 OMA PoC Functional Architecture

If we consider the set of functionality that a PoC service provides to an end user, we could draw a picture highlighting the high-level (and non-exhaustive) list of building blocks shown in Figure 2.2.

Regardless of the technology used to deliver a PoC service, the high-level functions demanded by a potential user to a generic PoC service can be grouped into:

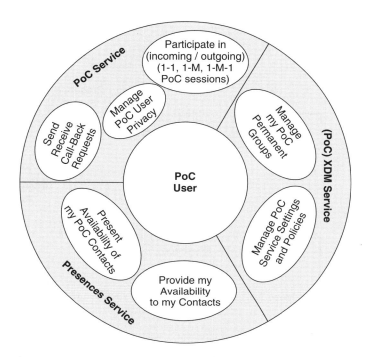

Figure 2.2   Overview of PoC functions from the PoC user perspective

a) specific PoC session functions,
b) capability to let the user manage PoC groups and service settings,
c) capability to provide dynamic 'availability' information about 'my PoC contacts'.

During the OMA PoC standardization process, end user requirements were defined, fine-tuned and described in a formal process, and were documented in [20]. After the requirements gathering process was reasonably mature – even though it would be finished significantly later – the architecture work started. During the architecture definition process, all the considerations and influencing factors described in section 2.2 were taken on board and, as an outcome of the whole activity, the OMA PoC architecture was finally designed and frozen in the *OMA PoC Architecture Document* [1]. This document defines a functional architecture that builds up the PoC ecosystem. Such architecture consists of PoC nodes (e.g. PoC Client, PoC Server, PoC XDMS) and their relationship to other enablers defined by OMA (XDM, Presence) or by 3GPP (e.g. SIP/IP Core), in order to build up an end-to-end service [1]. Figure 2.3 shows the OMA PoC standard architecture.

We can observe two types of elements in Figure 2.3:

• *Functional entities* defined by OMA Bold boxes refer to entities explicitly defined by the OMA PoC standard [1], while the rest of the boxes are defined in other OMA specifications (except the SIP/IP Core function, which is standardized by 3GPP). Each box identifies a

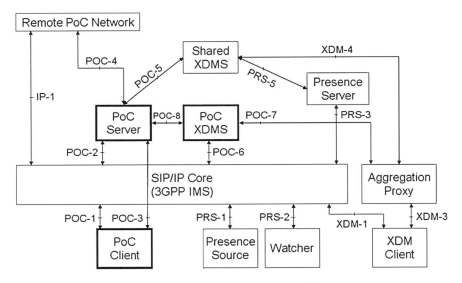

Figure 2.3   OMA PoC Functional Architecture

client or network function, which consists of a set of features and capabilities which can be logically grouped (e.g. a PoC client, a PoC server or a Presence server).

- *Reference points*, identified by an acronym followed by a number (e.g. POC-1). Whenever direct communication between two functions is required, OMA defines a reference point. A reference point consists in a well-defined set of operations that are used to communicate two logical functions. The complete definition of a reference point includes its mapping into a protocol (e.g. SIP, RTP, TBCP, XCAP), which is used to implement the operations required to fulfil the function of such an interface[2].

There are three PoC-specific functional elements defined in the PoC architecture (highlighted in bold boxes in the figure above): the PoC UE (client), the PoC server and the PoC XDMS (to store PoC groups and policies). The SIP/IP Core provides connectivity between clients and applications for SIP-based services (as is the case for PoC). Among other functions, the SIP/IP Core provides standard SIP registration and routing capabilities and, as such, is out of the scope of OMA standardization activities.

The rest of the elements in the architecture are used to support or enrich the PoC service: the Aggregation Proxy, the Shared XDMS and the XDM client are defined in the OMA XDM Enabler [6]. The Presence server, the Presence Source and the Presence Watcher are specified in the OMA Presence Enabler [7]).

Since OMA PoC is an open service with the capability to interconnect (and communicate) between several PoC networks, OMA also defines interfaces towards remote PoC servers and SIP/IP Core Networks (observe reference points IP-1 and POC-4 in Figure 2.3).

We can summarize an at-a-glance description of Figure 2.3:

---

[2] Although there is a slight technical difference between a *Reference Point* and a *Network Interface*, we will use both terms interchangeably. For additional details, the reader is referred to [12].

- The PoC Client and the PoC Server are the main elements supporting regular PoC operations (session setup and participation). In addition, the PoC XDMS stores XML documents containing PoC group or PoC policy definitions.
- The XDM elements (XDMC and Aggregation Proxy) support secure management operations into XML documents stored in any XDM server (e.g. PoC XDMS or Shared XDMS).
- The Presence Source and Watcher functions interoperate with the Presence Server to send and receive Presence information respectively.
- When needed, client-server or server-server communications make use of the underlying SIP/IP Core functionality that routes SIP signalling between SIP capable endpoints.

The logical architecture defined by OMA does not mandate how the different functionalities are mapped into physical entities. Therefore, it is feasible that different logical entities are grouped into a single physical node. Alternatively, a single logical function may spread across several physical entities. How and which elements are grouped together should be decided carefully, given that this election may influence into how network elements and clients – potentially supplied by different providers – should interoperate in an open environment based on the OMA standard.

The scope of OMA specifications covers all interfaces defining client-server communications, interaction of PoC with OMA XML Document Management (XDM) and Presence enablers, interoperability with external PoC networks to support communications across different domains, and usage of the SIP/IP core platform (e.g. IMS). The client, in turn, supports communication with the OMA PoC, Presence and XDM services over the SIP/IP core via the corresponding client-server interfaces.

In order to support all PoC, Presence and XDM operations, OMA defines a set of eight PoC-specific reference points (POC-1 to POC-8), 15 Presence-specific reference points (PRS-1 to PRS-15) and four XDM-specific reference points (XDM-1 to XDM-4) – not all Presence and XDM reference points are shown in Figure 2.3. Furthermore, the IP-1 reference point implemented by the SIP/IP Core supports network-to-network communication for SIP-based services such as PoC.

The next sections will briefly describe the functionalities of each element involved in end-to-end PoC communications. XDM and Presence functions are described in detail in chapters 3 and 4 respectively.

2.4 PoC Client

2.4.1 Introduction

The PoC client resides in the mobile terminal and lets the user access the PoC service. It supports all signalling, media and talk burst control procedures required to interact with the network entities (e.g. SIP/IP core and PoC server). In general, the PoC-enabled device will implement PoC client and XDM client features, while it may support Presence features as well (e.g. Presence Watcher and Presence Source), in which case it also displays a Presence-enabled PoC contact list.

Effectively, the PoC client provides the interface that connects the high-level functions described in Figure 2.2 with the end user.

We can divide PoC client functionalities into *registration*, *session related*, *non-session related* and *service configuration* procedures. We briefly describe them in the sections below.

2.4.2 Service Registration Functions

The PoC client shall be able to authenticate and perform registration of the PoC user against the SIP/IP Core. Successful registration is a prerequisite in order to be reachable by other users, or to start PoC sessions upon user request.

After registering with the SIP/IP Core, the PoC client typically contacts the PoC server to configure its PoC service settings, such as the PoC answer mode (automatic/manual) or its ability to support simultaneous sessions. This process is called *Publication of PoC Service Settings* [8] because it makes use of the SIP PUBLISH transaction [9].

During the registration stage, the PoC-enabled UE may also contact the Presence server to inform it about its new status and availability (e.g. *I am now registered and available for PoC communications*). It is important to differentiate among SIP registration (if a user is unregistered, she will not receive any incoming call request: the user is unreachable from the SIP signalling perspective) and publication of Presence information: Presence is used to actually inform other subscribers that a given user has become available.

Instead of the PoC handset proactively publishing Presence information on behalf of the user, OMA PoC supports an optional feature: when the PoC service detects that a user has become online, it may publish this information towards the Presence server. This feature allows for a simpler PoC client implementation and saves some bandwidth in the wireless interface. For additional details on this feature, the reader is referred to chapter 4.

2.4.3 Session-Related Functions

The key functionalities that the PoC client must support, related to PoC sessions are:

- Initiate PoC sessions, support session participation (e.g. talk or listen) and leave sessions upon user request or when instructed by the PoC server.
- Process incoming call requests according to the configured service settings (e.g. accept automatically or prompt the user to manually accept or reject incoming PoC sessions).
- Support talk burst control procedures (i.e. procedures to negotiate who has the right to speak at any time) and Talk Burst Control Protocol (TBCP) negotiation[3].
- When the user has the right to speak, generate and send talk bursts.
- When the user is in listening mode, receive talk bursts, decode them, generate corresponding audio output and display information about the talking user (if available).
- Support User Plane adaptation procedures. This feature is generally invoked by the PoC server when a new user joins a group session and, for some reason, the media parameters have to be renegotiated among all participants (e.g. the new user does not support the CODEC settings currently in use for the session). In this case, all participating clients must adapt to the new media settings. These procedures may include modifying the selected CODEC (e.g. from AMR-WB to AMR-NB), the CODEC rate or modifying the CODEC framing options used to generate RTP packets.

---

[3] Although OMA PoCv1 theoretically supports negotiation of which protocol will be used for floor control procedures, there is only one protocol actually defined and supported: TBCP. This feature may have some applicability in OMA PoCv2, which is being standardized at the time of writing.

OMA also defines a set of PoC client optional requirements. Some of them are relevant in specific environments, while others allow some performance optimizations which improve user experience in certain cases. A high-level overview of these requirements follows:

- Support multiple Talk Burst Control Protocols. Although theoretically feasible, OMA PoCv1 clients will only support the TBCP defined by OMA [4].
- Support Talk Burst Request queuing. When Talk Burst Request queuing is supported, the PoC server may queue TBCP messages from one or more users requesting the right to speak. This feature lets users request the floor while others are still talking. When this feature is not available, the default policy is to assign the right to speak only when the session is idle (e.g. when a participant has stopped talking), and only to the first user requesting the floor.
- Send RTCP-based quality feedback reports after the end of a talk burst.
- Support Pre-established Sessions. This feature allows a PoC client to establish a 'virtual' session with the PoC server, so that when a 'real' call is to be initiated later on, the signalling procedure is shortened. This feature reduces session setup delay.
- Simultaneous Sessions. When this feature is supported, a PoC client may simultaneously participate in two or more PoC sessions. This feature may lead to lost talk bursts when two streams pertaining to different PoC sessions collide (the PoC server will ony deliver one of them and discard the other one).
- Support Session On-Hold procedures. When a user does not want to receive talk bursts from other participants, they may place media 'on hold'. The user is still able to send talk bursts (when allowed by the server). Talk burst reception can be re-activated by putting media 'off-hold'.
- Support User Privacy. When the user has activated Privacy, the PoC client requests that user identity is hidden from other participants during a PoC session. This feature lets a PoC user send invitations to participate in a PoC session hiding her PoC identity (similar to the very familiar *Calling Line Identification Restriction* – CLIR – GSM supplementary service). In turn, once the session is established, when a user with privacy activated is talking her identity is not shown to the rest of participants[4].

2.4.4 Non-Session Related Requirements

Apart from attaching to a SIP/IP Core and participating in PoC sessions, there are several non-session related functions defined for a PoC client, namely:

- Support reception of *Instant Personal Alert* (IPA) messages. A PoC user may send an IPA message to another PoC user to request that he/she calls him back (using the PoC service). This is generally used when the recipient is known to be busy or it is not convenient to start a PoC session. Instant Personal Alert is a transaction based on the SIP MESSAGE request [10]. Therefore, it does not generate a SIP session (although the recipient may start a PoC session later on).

---

[4] As one would expect, the network entities such as the SIP/IP Core and the PoC Server are aware of each user's identity, regardless of their associated Privacy setting.

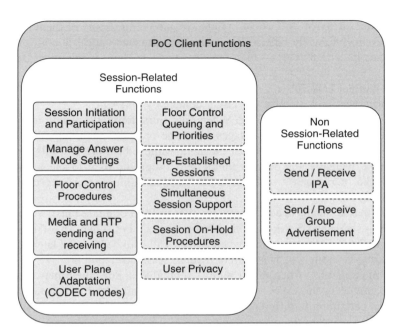

Figure 2.4 High level overview of PoC Client functions

Being able to receive IPA messages is a mandatory client PoC feature. However, sending Instant Personal Alert messages is an optional client feature. Handsets not supporting this feature would not offer any option to send IPA messages in the PoC client user interface. Correctly forwarding (i.e. receiving and sending) IPA messages is a mandatory PoC server feature.

• Group Advertisement (GA) support. This feature lets a user create a PoC group and provide information about the newly created group (and its associated PoC identity) to all members of the group. This way, members become aware of the existence of the group and can start group sessions relating to it, if they wish. This feature is also based on the SIP MESSAGE transaction.

Unlike IPA, the ability to send or receive Group Advertisement messages is optional for PoC clients and servers. PoC servers must however be able to correctly process incoming GA messages when this feature is not supported (i.e. deliver a SIP error message to correctly terminate the transaction and indicate that it could not be fulfilled).

In order to provide an at-a-glance overview of PoC client functions, Figure 2.4 presents an overview of session-related and session-unrelated client features. Capabilities in dashed boxes are optional [8].

2.4.5 Service Configuration Requirements

In order to support consistent service configuration, the OMA Device Management (DM) service supports remote client configuration of default PoC settings and GPRS parameters [11].

In addition, the PoC client must be able to set certain service parameters according to user preferences. These settings include:

- *Answer Mode Indication* (Manual Answer, Automatic Answer). This parameter governs whether incoming sessions will be automatically accepted or if explicit user acceptance is required.
- *Incoming PoC Session Barring* (ISB) and *Incoming Instant Personal Alert Barring* (IAB). These features let the user temporarily block incoming sessions and incoming Call-Back requests (e.g. if the user enters an important meeting she may activate these features in order not to be disturbed by PoC activity).
- *Simultaneous PoC Sessions Support.* This setting lets a user indicate whether he/she supports and is willing to use Simultaneous PoC sessions. The user is also able to indicate the maximum number of simultaneous PoC sessions. When the maximum number of simultaneous sessions is reached, new incoming sessions will be rejected.

2.5 XML Document Management Client

The XML Document Management Client (XDMC) communicates with the XDM servers (via the Aggregation Proxy) located at the network. This lets the user create, modify, retrieve and delete XML documents. The main purpose of these documents is to define groups (which later on will be used to set up PoC group sessions) or to specify PoC service policies and rules. A typical example is a PoC rule defining a blacklist. Whenever a blacklisted PoC user tries to reach the owner of the blacklist document, the call will be automatically rejected by the PoC server (without even alerting the called party).

XML Documents may be PoC specific or general ones which can be shared with other applications. As an example, a list of identities may be reused by PoC and Presence to define a service-specific document such as a PoC group or a Presence list. Shared documents are stored in the Shared XDMS, which can be accessed by any other XDM server.

IETF has defined a mechanism aimed at specifying management operations in XML documents between the XDMC and any XDMS. This mechanism is the *XML Capability Access Protocol* (XCAP) [6]. XCAP runs on top of HTTP or HTTPs. In the particular case of OMA XDMC, the XDMC uses XCAP to contact the Aggregation Proxy. The Aggregation Proxy performs management operations on the XDMS nodes on behalf of the authenticated XDMC.

Interaction between the XDMC and the Aggregation Proxy and the different types of XML documents managed by the XDMC are described in chapter 3.

2.6 PoC Server

2.6.1 Introduction

The PoC server is the main network entity providing services to PoC clients. It supports SIP, RTP, TBCP, RTCP and XCAP protocol implementations to interact with PoC clients, other PoC servers, external enablers such as XDM servers, and communication with and via the SIP/IP Core.

In general, the PoC server is responsible for establishing PoC sessions upon client request, terminating sessions when certain conditions are met, handling floor control procedures to determine which user has the right to speak at a given time, replicating and forwarding media packets to session participants, delivering Instant Personal Alerts and Group Advertising messages. The PoC server supports additional back-end functions such as logging, user charging, provisioning or service monitoring and management.

In order to enable service scalability and implement PoC communications across several networks, OMA has defined two different PoC server roles, namely: the *Controlling PoC function* and the *Participating PoC function*.

The idea is that when a PoC session involves communication across different PoC domains, or communication of users within a large domain with several deployed PoC servers, communication among PoC servers is required to connect all participants in the session. In such cases, one server takes the leading 'Controlling' role, while the rest of the servers assume the 'Participating' function.

2.6.2 Split of Functionalities between the Controlling and the Participating PoC Functions

As explained in the previous section, a PoC session requires involvement from one single Controlling PoC function and one or more Participating PoC functions. The Controlling PoC function provides centralized management of the conference, while the Participating PoC function supports management and policy functions for clients under its service. This clear split of functionalities and the definition of how Controlling and Participating PoC functions interact during a session, let the PoC service extend through several network domains, enabling communications between users that are served by different operators in a clearly defined and scalable way.

The PoC server that provides services to a user starting a new 1-to-1 or Ad-Hoc group PoC session is the one performing the Controlling PoC function role. Let us consider the following example:

1. John is a PoC subscriber at wirelessfuture.com, with PoC address: sip:john@ wirelessfuture.com
2. Alice and Mark are PoC subscribers of otherdomain.com, with PoC addresses: sip:alice@ otherdomain.com and mark@otherdomain.com
3. John calls Alice and Mark.

In this case, the PoC server at wirelessfuture.com will perform the Controlling PoC Function for the whole duration of the session.

For Pre-arranged group sessions or Chat sessions the PoC server that hosts the group will be the one performing the Controlling PoC function. Once this role is assigned, it remains unchanged until the end of the session. As an example, if John calls a Pre-arranged PoC group called sip:alice_colleagues@otherdomain.com, the PoC server at otherdomain.com will perform the Controlling PoC function, regardless of an external user (sip:john@wirelessfuture.com) having initiated the session.

The logical split between Controlling and Participating PoC functions applies as well to intra-operator PoC sessions. However, in this case this logical split could actually be implemented in a single physical entity. On the other hand, in case of distributed multi-server deployments where several servers are required to support a large customer base, the distinction between Controlling and Participating PoC Function generally maps to physical split as well.

Figure 2.5 shows an example where a Pre-Arranged Group session between Mary, Joe and Charlie (all of them served by different PoC providers) eventually leads to a Controlling PoC function that does not directly handle any PoC user (as Charlie switchess off his PoC device shortly after the call was established). In this case, Joe and Mary may keep talking through the interconnection of their corresponding Participating PoC functions and the Controlling PoC function that hosts the session.

The Controlling PoC function is responsible for centrally managing all SIP signalling, TBCP floor control and RTP media replication and forwarding towards the participants in a session. In particular, the Controlling function centralizes all floor control operations (i.e. deciding who can speak). This setup significantly simplifies the PoC service design, since it would have been very complex to establish an architecture where floor control decisions are made in a distributed fashion.

When it comes to media management, the fact that the Controlling PoC function performs media replication (all RTP packets sent by a talking PoC user must be replicated and distributed to the rest of participants in the session) leads to a simpler design, at the cost of additional overhead when interconnecting PoC domains. Effectively, when a participant from network X is talking to N participants in network Y, N RTP packets must be generated by the Controlling PoC server (assuming it is located in network X) and delivered to network Y, to be forwarded to all N recipients. This situation, however, is inherent to the usage of SIP signalling, which requires a media stream per each SIP dialog established between domains.

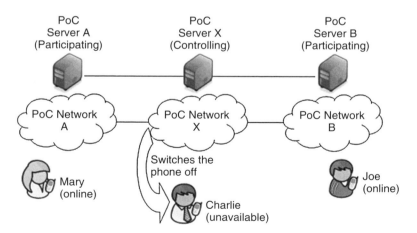

Figure 2.5   In this example, the Controlling PoC function will handle a session where all participants belong to remote PoC networks

The Participating PoC function supports several functionalities such as session initiation on behalf of the user, user charging or privacy support. The Participating PoC function is always in the SIP signalling path between end users and the Controlling PoC function, while it may optionally be in the media (RTP) and floor control (TBCP) path as well.

In general, exchange of SIP signalling between PoC clients and PoC servers and between PoC servers takes place through one or more IMS core networks, which manage routing of SIP messages across domains.

Table 2.1 provides a summary overview of the split of functionalities between the Controlling and Participating roles. The next sections further detail them.

Whenever the conventions *C-PoC* and *P-PoC* are used, they refer to the Controlling PoC function and the Participating PoC function respectively.

2.6.3 Controlling PoC Function

The Controlling PoC function is responsible for centrally managing signalling, talk burst and floor control procedures. Due to this centralized management approach, it is possible to avoid floor request conflicts and to easily manage incorporation of new users to a PoC session. The Controlling PoC function capabilities are further described below.

2.6.3.1 Session Handling Functions

The Controlling PoC function centrally manages all PoC session related signalling, media and floor control procedures. It also implements policy enforcement mechanisms for participation in group sessions (e.g. ensuring that the maximum allowed number of participants in a group session is not exceeded).

2.6.3.2 SIP Session Control Functionalities

The Controlling PoC function is responsible for centrally managing SIP session origination, release, users joining and rejoining PoC sessions and delivery of Instant Personal Alert and Group Advertisement messages.

The Controlling entity may also support the ability to notify users about the state of a PoC call and, in particular, about the participants involved in that session. This feature lets all participants in a session receive dynamic information about it (who is participating, how many listeners are connected, who is talking, . . .). The only entity in the end-to-end path that has a general overview of the session is the Controlling PoC function, which is able to distribute this information dynamically towards all participants interested in knowing the status of the conference.

The Controlling PoC server must also support the ability to negotiate the Talk Burst Control Protocol to be used during the subsequent PoC session. However, OMA has only defined support for the OMA-defined TBCP for floor control, therefore converting it into the only available option.

Table 2.1 Summary of Participating and Controlling PoC functions

Controlling PoC Function (C-PoC)	Participating PoC Function (P-PoC)	Comments
Centralized PoC session handling	PoC session handling	
Centralized media replication and distribution	Relay RTP media packets between the C-PoC	The P-PoC shall perform this function when it remains in the media path between the PoC client and the C-PoC
Centralized Talk Burst Control functionality including Talker Identification	Provide the Talk Burst Control message transfer function between PoC client and the C-PoC	The P-PoC performs the indicated function only when it remains in the media path
SIP Session handling, such as SIP Session origination, release, etc.	SIP Session handling, such as SIP Session origination, release, etc, on behalf of the represented PoC client	
Supports policy enforcement for participation in Group Sessions	Provides policy enforcement for incoming PoC session (e.g. Access Control, Incoming PoC Session Barring, availability status, etc)	For example: controlling that the maximum allowed number of participants is not exceeded
Provides Participants information	–	
Supports privacy of the PoC Addresses of Participants who have enabled Privacy	Supports privacy of the PoC Address of the Inviting PoC User during PoC session setup in the terminating PoC network	That is: the P-PoC removes the identity of the calling user to the called one
Collects and provides centralized media quality information	Collects and provides media quality information	The P-PoC shall perform this function when it is in the media path between the PoC client and the C-PoC
Provides centralized charging reports. Supports PoC session owner charging	Provides Participant charging reports. Supports PoC session participant charging	The P-PoC is the one charging participants, while the C-PoC may charge the session owner
User Plane adaptation procedures	User Plane adaptation procedures	
Talk Burst Control Protocol negotiation	Talk Burst Control Protocol negotiation	Generally not implemented as the only defined Talk Burst Control Protocol is TBCP
–	Stores the current Answer Mode, Incoming PoC Session Barring and Incoming Instant Personal Barring preferences of the PoC client	The P-PoC is the one serving its local user. Therefore, it is the one aware of local PoC service settings of each user
–	Filtering of the media streams in the case of Simultaneous PoC Sessions	
Transcoding between different codecs.	Transcoding between different codecs.	This functionality is optional. The P-PoC may support it only when it is present in the media path

2.6.3.3 Media and Talk Burst Control functionalities

Given the half-duplex nature of the service, one of the key responsibilities of the Controlling PoC function is to centrally assign and arbitrate the right to speak among the different users involved in a call. TBCP is used for that purpose. When assigning the right to speak, the PoC server also provides indication of the identity of the current talker.

Another key role implemented centrally is RTP media packet *replication* and forwarding towards all active receivers involved in a session.

The Controlling PoC entity must perform these functions without prejudice on the Privacy settings that have been activated by each participant in the call: it must not propagate the talker identity in TBCP packets if Privacy is activated for the talking user.

The Controlling PoC server may also implement the *User Plane Adaptation* feature. The goal of this feature is to ensure that the CODEC configuration selected for a session interoperates with all clients involved in that session. For example, if a user with a reduced set of available CODEC modes joins an ongoing session, this option may be used to *downgrade* some participants and enable seamless communication between all participants.

Additionally, the Controlling server must support media renegotiation when explicitly initiated by a PoC client [4]. A use case for this feature would be a client roaming from 3G to a 2G coverage area and thus wishing to move from a high bandwidth AMR mode towards a lower one.

Negotiation of *User Plane Adaptation* procedures is implemented via SIP signalling, by reissuing a SIP INVITE message with a new SDP offer detailing the new media parameters applicable to the session.

Finally, the Controlling entity may support transcoding between different audio CODECs if required (e.g. transcoding from/to AMR to/from EVRC when 3GPP and 3GPP2 devices are involved in the same session).

In Figure 2.6 a high-level overview of the Controlling PoC function managing TBCP and RTP flow is presented. Observe how the Controlling PoC function assigns the right to speak,

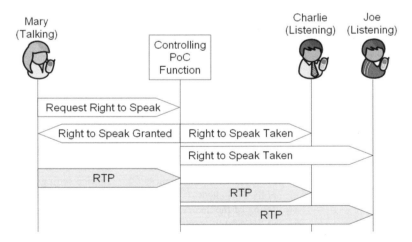

Figure 2.6   Controlling PoC function managing TBCP and RTP flows

and how it is in charge of replicating RTP packets towards all participants in a PoC session. The corresponding Participating PoC functions for Mary, Charlie and Joe have been ommitted from Figure 2.6 for sake of clarity.

2.6.3.4 Charging

In general the Controlling PoC function does not support direct user charging, because it does not necessarily serve any user directly (users are handled by the Participating PoC Function).

However, the Controlling PoC function is capable of charging the PoC session owner. Effectively, a PoC user may initiate a PoC session (e.g. Pre-Arranged) and he/she may later on leave such session. In that case, no Participating function serves the session owner anymore, as the owner is no longer involved in the session. If the service provider wishes to charge the session owner based on the duration of the session (or any other session-related parameter), regardless of her being actively involved in the session, the Controlling PoC function may support such feature.

Additionally, the Controlling PoC function may charge the owner of a PoC group. Consider that a PoC user creates a Chat PoC group called sip:gossips@wirelessfuture.com. The service provider may charge the owner of that group based on the activity of the group, regardless of the group owner (the one who created it) participating in those chat sessions or not.

2.6.4 Participating PoC Function

The Participating PoC function is involved always in the signalling path between the client and the Controlling function. Additionally, the Participating PoC function may remain the media (RTP) and the floor control (TBCP) paths as well. However, there is the option that RTP and TBCP streams are exchanged end-to-end between PoC clients and the Controlling PoC function.

Given that the Participating PoC function is generally in charge of providing services to its local customers, it generally applies policy enforcement to ensure that the client behaves according to operator policies (e.g. perform credit check before forwarding requests towards the Controlling entity). For this reason, it is expected that the Participating PoC function will generally remain in the RTP/TBCP path (*Participating Function in the Transport Path* option, as defined in the PoC User Plane specification [4]).

The following sections provide further details about the standard functions that the Participating PoC server implements.

2.6.4.1 PoC Service and Session Handling Functionalities

The Participating PoC Sever handles the sessions between the clients it serves and the Controlling function. It proxies SIP signalling, such as session origination, release, etc. on behalf of the represented PoC clients. The Participating function stays in the signalling path for all SIP messages exchanged between the PoC client and the Controlling function.

An important role of the Participating server is to implement policy enforcement to govern how local clients interact with the service. In particular, the Participating PoC function should check the following information before delivering an incoming SIP message to a served PoC user:

1. Evaluate whether the calling user is included in a PoC black-list stored in the PoC XDMS under the folder of the called user
2. Evaluate availability status for the called user
3. Evaluate the Incoming Session Barring (ISB) flag status (if ISB is activated, incoming sessions must be automatically rejected).
4. Evaluate the Incoming Alert Barring (IAB) status (if IAB is activated, incoming PoC Alert messages must be automatically rejected).

In order to perform these tasks, the P-PoC must store PoC settings such as the current Answer Mode, Incoming PoC Session Barring and Incoming Instant Personal Barring preferences of the PoC client. Such settings are activated by the PoC client every time it is activated (by sending a SIP PUBLISH message towards the PoC server assigned to the user).

Finally, the Participating node is responsible for ensuring privacy of the address of the inviting user against the terminating network, when the user has activated it. Thus, the Participating function removes the originating user PoC address from the SIP INVITE message delivered to an invited PoC user under its domain.

2.6.4.2 Media and Talk Burst Control Functionalities

In general, during session setup the Participating PoC server provides the proper SDP parameters to stay in the media and TBCP path between the Controlling and the client entities. However, if the Participating PoC entity does not have any policy that requires involvement in such flows, it may effectively remove itself from the media and floor control paths.

When present, the Participating PoC function supports relay of RTP packets and message transfer of TBCP packets between the PoC client and the Controlling PoC server. It must also collect and provide media quality information based on RTCP reports towards the two communicating endpoints.

When the Participating PoC function is in the media path it may also support CODEC transcoding, media filtering when a user is involved in two or more simultaneous sessions and Talk Burst Control Protocol translation,although this functionality is, in general, not required for the OMA PoCv1 Enabler:

2.6.4.3 Charging

The Participating PoC function is responsible for charging all PoC users under its control, thus supporting different charging paradigms, depending on service provider's policies. If a service provider implements a charging model that requires information about session activities, the Participating PoC function must provide such information to the charging system. Example charging models include charging based on sent talk bursts, duration of sent talk bursts or initiation of PoC sessions. Whenever any of these charging models is implemented, the P-PoC must stay in the media and floor control paths, so it must indicate such configuration in SIP/SDP headers and parameters during session setup, as indicated in the previous paragraph.

Observe that a flat-fee billing model (e.g. monthly fee without usage limit) may not require direct involvement of the Participating PoC function when a user is engaged in a session. In this case, the P-PoC may safely remain outside of the media and floor control paths, because the service revenue is totally decoupled from the service usage. The P-PoC function may perform a credit check operation for the called user whenever a SIP INVITE message is received and no further communication with the Charging System is required for the duration of that session.

2.6.5 User Plane Routing Configurations between P-PoC and C-PoC Functions

Now that the roles of the Participating and Controlling PoC Functions have been presented, we will introduce the two different options that may be configured to enable communication between them in the scope of a session. The two alternative configurations are: *Participating PoC function in the Transport Path* and the *Direct media flow between Controlling PoC function and PoC client* [1].

2.6.5.1 Participating PoC Function in the Transport Path

It is expected that, in general, the Participating PoC function will be configured to stay in the transport path (i.e. dealing with RTP, RTCP and TBCP flows). This configuration is shown in Figure 2.7.

In this configuration, the P-PoC inserts its own IP address and UDP ports in the SDP message of the SIP request (e.g. INVITE) in order to stay in the media (RTP), talk burst control (TBCP) and quality feedback (RTCP) paths. In this case, the P-PoC behaves as a so-called SIP *Back-to-Back User Agent* (B2BUA). It will look to the C-PoC entity as another user agent that participates in the session on behalf of the end user.

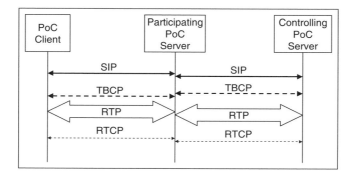

Figure 2.7   Participating PoC Function in the Transport Path

The Participating function may stay in the media path because of a number of reasons; the following list briefly outlines some of them:

- The PoC session reuses an already existing pre-established session, which was already setup beteen the end user and the P-PoC entity.
- The PoC client and the PoC server support the Simultaneous Sessions feature. In this case, the P-PoC must stay in the media path to perform the filtering function associated to Simultaneous Sessions (i.e. to decide which talk burst must be delivered to the PoC client if two talk bursts from different sessions collide).
- The charging model implemented by the operator providing the PoC service requires information from the transport layer (refer to chapter 5 for detailed information about PoC charging).
- The Participating entity performs CODEC translation (e.g. from/to AMR to/from EVRC).
- The Participating entity performs talk burst control protocol translation. This function will never be met for OMA PoC communications, although it may be required when the P-PoC interconnects an OMA PoC network with a legacy P2T system using another protocol. In the future, other floor control protocols (e.g. Binary Floor Control Protocol, BFCP [12] may become available, and this feature may gain more importance.
- The operator has decided to insert the P-PoC entity in the media path for local policy reasons.

In case the above reasons are not relevant for a given deployment, the P-PoC may remove itself from the media path, as described below.

2.6.5.2 Direct Media Flow between Controlling PoC Function and PoC Client

In this configuration RTP media, RTCP feedback and TBCP floor control messages flow directly between the PoC Client and the Controlling entity, as shown in Figure 2.8. In this case the IP addresses and ports notified in the SDP offer answer issued by the PoC Client are transparently delivered towards the C-PoC.

Some advantages of this configuration are:

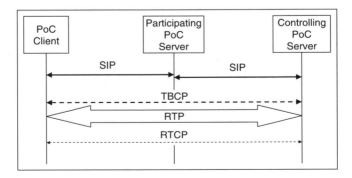

Figure 2.8 Direct media flow between Controlling PoC Function and PoC Client

- Lower processing delays at the home network of the P-PoC function. Since RTP, RTCP and TBCP packets need not be evaluated up to the application layer (e.g. such as in the previous case) lower delays may be expected for this configuration.
- Minimized hardware and software requirements for the P-PoC function implementation. Since the P-PoC is only involved in SIP signalling processing, it does not need to allocate UDP ports, processing power and network resources to manage RTP, RTCP and TBCP traffic.

When the P-PoC server is configured not to be involved in the media path, it manages SIP signalling exchange in *SIP Proxy* mode (not in B2BUA mode). The P-PoC server will ensure that it stays in the signalling path for the duration of the session (by inserting the Record-Route SIP header in the INVITE request), so that it receives all relevant SIP messages exchanged between the client and the C-PoC.

However, configuration of the P-PoC in *SIP Proxy* mode means that the P-PoC cannot proactively send any SIP message to the PoC client or the C-PoC (it can only *proxy* messages received either from the client or from the Controlling entity). Therefore, it is not possible for the P-PoC to apply policies to perform session management functions (e.g. terminate a call leg when a user has run out of credit). For this reason, it is expected that this configuration will only be used when the charging model allows.

It is worth observing that in environments where several P-PoC entities are involved (e.g. a call with users from several different PoC networks) the two configurations may coexist in the same session. In Figure 2.9 P-PoC A is configured to stay in the media and transport path, while for client B the RTP, RTCP and TBCP flows are routed directly between the C-PoC-X and the client (i.e. P-PoC-B does not stay in the media path).

We could imagine that the operator in charge of PoC Server A wishes to charge client A based on actual usage (e.g. number or duration of the talk bursts sent by user A), while the operator in charge of PoC Server B uses an access fee model which does not require usage monitoring to charge customer B.

Observe that regardless of the P-PoC remaining or removing itself from the media plane routing, from the SIP perspective, PoC Clients are always connected to their P-PoC function. The Participating PoC function relays SIP signalling towards the Controlling function in B2BUA or Proxy mode.

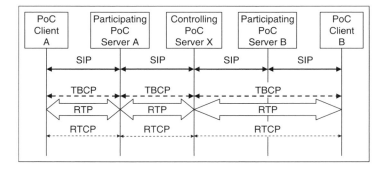

Figure 2.9   PoC session with P-PoC A in the media path and direct media flow towards client B

2.6.6 Interaction with Group and Policy Management Entities

In order to support group-based communications, OMA has defined the XDM architecture, which manages XML documents created by users.

These XML documents are used to define groups and service policies. In the particular case of OMA PoC, the PoC server supports management of PoC-specific XML documents from the PoC XML Document Management Server (PoC XDMS), and retrieval of shared documents from the Shared XDMS.

PoC-specific XML documents may point to external lists of contacts (e.g. URI lists) which are stored in the Shared XDMS. The goal of this split of information is to enable sharing of user created groups (URI lists) across several services (e.g. PoC, Presence, Conferencing). Service-specific documents (e.g. a PoC group stored in the PoC XDMS) contain external references that point to shared documents, and define how a document stored in the Shared XDMS is used in scope of a particular service (e.g. PoC).

2.6.7 Interaction with the OMA Presence Enabler

The PoC server may optionally behave as a Presence Source to communicate with the OMA Presence Enabler on behalf of the user. This feature enables some traffic optimization over the air: effectively, when a user logs in to the PoC service (i.e. it has published her PoC service settings), the server may proactively publish user status as being *available for PoC communications* in the Presence service.

This feature of the PoC server avoids the need for the PoC-enabled UE to explicitly communicate with the Presence Enabler after registering to the SIP/IP Core. In this case, the PoC UE may not need to implement the Presence Source function, since all Presence-related publications can be managed by the PoC Service. Obviously, this feature only supports limited functionality: if the user wishes to publish complex Presence information, such as *I am currently in a meeting, so not available for PoC sessions, but available for instant messaging and video sharing*, a dedicated Presence Source function at the UE may be required.

Additionally, the PoC server may also behave as a Presence Watcher and subscribe to Presence information from a user. This mechanism would let the PoC server support policy enforcement based on Presence information (e.g. if a user is 'not-available' according to the Presence service, then the PoC server will not try to contact the user when an incoming call arrives). Of course, making use of this feature requires the assumption that the Presence service information is 'reasonably' accurate and timely.

2.7 PoC XML Document Management Server

The PoC XML Document Management Server (PoC XDMS) is defined by the OMA PoC architecture as the element responsible for storing and managing PoC-specific XML documents.

Users may create, retrieve, modify or delete these documents and their contents. PoC-specific XML documents are used to define permanent groups and PoC-specific policies to be applied when setting up PoC sessions (e.g. a user may store an access control list in the PoC XDMS to control which users are allowed to invite him/her).

The PoC XDMS supports communication with the PoC server to deliver PoC-specific XML documents. These are then used by the server to apply the defined policies or use the group information to setup a group-based PoC session.

The PoC XDMS and the PoC server are different logical entities defined by OMA. However, commercial solutions may actually integrate the PoC server and PoC XDMS functions in the same entity, thus avoiding the need to build up the PoC – PoC XDMS XCAP interface (POC-8). This situation, however, should evolve towards a fully decoupled solution, in order to ensure a future-proof architecture.

Further information about general OMA XDM features is provided in section 2.8.3 and in chapter 3.

2.8 External Entities Providing Services to PoC System

In addition to the PoC-specific elements and functionalities described in the previous sections, OMA references a set of external entities required in order to deploy a complete end-to-end PoC service that fulfills all functions outlined in Figure 2.2.

These entities are, namely: the underlying Access Network and SIP/IP Core (e.g. IMS), the OMA XDM and Presence Enablers, an external charging entity to enable PoC charging and, finally, the OMA Device Management entity to support remote configuration of PoC-enabled devices.

These elements are briefly presented in the next sections.

2.8.1 Access Network

As an IP-based service, PoC requires end-to-end connectivity between the different endpoints involved in the communication (e.g. PoC client, IMS, PoC server). Given its low bandwidth requirements[5], PoC may be run on top of several wireless technologies, including 2.5G (e.g. GPRS, EDGE), 3G (e.g. WCDMA, cdma2000), 3.5G (e.g. HSDPA, EUL) and Wi-Fi (e.g. IEEE 802.11x). When relevant, the focus of this book will be put on PoC technology deployed over 3GPP-enabled networks such as GPRS, EDGE, WCDMA or HSDPA. In any case, most of the service-related aspects are independent of the technology used to access the service.

2.8.2 SIP/IP Core (IMS)

PoC is the first OMA approved SIP-based service to be deployed over 3G wireless networks. Even though OMA PoC is access network agnostic, it is expected that the SIP/IP Core functionality shown in Figure 2.1 will be delivered by the architecture defined to support multimedia services in 3G networks: the IP Multimedia Subsystem (IMS) standardized by 3GPP or the Multimedia Domain (MMD) as standardized by 3GPP2.

When OMA started PoC activities in mid-2003, standardization work in 3GPP around the IMS concept and architecture was already consolidated (similar work took place in 3GPP2

---

[5]Some flavours of the AMR CODEC may work at a rate as low as 4.75 kbps – excluding RTP/UDP/IP/L2 headers.

with the MMD architecture). Hence, OMA took 3GPP IMS as the natural platform to support SIP/IP Core features over cellular networks. Later on, a split of responsibilities between 3GPP/3GPP2 and OMA was defined. After solving some overlapping between both organizations, a general agreement was eventually reached and documented in [14][15].

As a rule of thumb, we can consider that general (non-service specific) SIP features and functions highly dependant on core and IP networking technology are managed by 3GPP, and should fall in the IMS area. This way, applications can concentrate on deploying consistent service logic, without needing to be involved in common or mobile-specific issues. On top of this general-purpose technology, OMA defines services and service enablers and focuses on the application and service layer definition.

Among others, IMS provides features such as user authentication, authorization and registration, user privacy at the control plane, SIP message routing, support for network interconnection or lawful interception capabilities.

Observe that, even though OMA and 3GPP ensured clear split and smooth interoperability between PoC and IMS, OMA PoC is still SIP/IP Core agnostic: assuming that a set of features is supported by the underlying SIP/IP Core, OMA PoC can be deployed over non-IMS infrastructure as well. In 3GPP networks, though, the natural choice to fulfil this functionality is – as explained above – the IMS/MMD. Hence, both 'SIP/IP Core' and 'IMS' will be used through the text interchangeably.

Detailed description of how OMA PoC interoperates with IMS will be presented in chapter 5.

2.8.3 XML Document Management Entities

The XDM Enabler defined by OMA [6] specifies a general architecture to manage XML documents. XML documents support service-specific configurations, policies or settings. In the particular case of PoC, the PoC XDMS supports two main services, namely: group management and definition of service policies. Two example use cases are:

- A user may create permanent PoC groups that are used by the PoC server to establish group sessions;
- A user may configure PoC access lists to determine which other users are allowed or blocked when trying to communicate with him/her.

These two capabilities are integral elements of any PoC service. Therefore, support for XDM operations is mandatory for any OMA PoC deployment.

Apart from the PoC XDMS, the two main elements of the XDM architecture that enable group and policy management operations are the Shared XML Document Management Server (Shared XDMS) and the Aggregation Proxy [6]. These are briefly described below.

2.8.3.1 Shared XML Document Management Server (XDMS)

Like the PoC XDMS, the Shared XDMS is an XCAP Server that manages XML documents. The main difference between the PoC XDMS and Shared XDMS is that the Shared XDMS stores lists of contacts (so-called *URI lists*), which are shared between several applications (e.g. PoC, Presence, IM or future to-be-defined services). In particular, the Shared XDMS does not store any enabler-specific information. Consider the following example:

- Alice wishes to create a PoC blacklist (Figure 2.10), because she has been recently spammed by some specific users, and she does not want to be bothered again. Additionally, she realizes that she would not like these contacts receiving Presence information about her.
- Alice uses her XDM client to create a new document in the Shared XDMS. The XDMC stores the list of users to be blacklisted in that document.
- Finally, Alice's XDMC creates a PoC policy and a Presence policy document, and stores them in the PoC XDMS and the Presence XDMS respectively. Both documents contain a pointer to the original document stored in the Shared XDMS.
- Whenever a change is made to the list of blacklisted users in the Shared XDMS, this change will be automatically ported to the PoC and Presence policies documents, without the need to perform any explicit change on those documents.

The next figure provides a high-level overview of this use case, where Alice's XDM client first (1) creates a list of users to be blacklisted, and stores it in the Shared XDMS, and then (2) it creates a two service-specific policy documents (blacklists), whith pointers to the shared list.

Observe that in general the Shared XDMS is only relevant when the user wishes to share XML documents across several enablers. Otherwise, the PoC XDMS is the right place to store all information that is not required by other applications.

The Shared XDMS supports creation, modification, retrieval, and modification of XML documents.

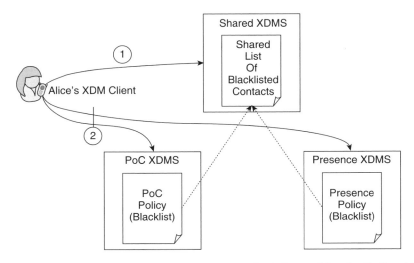

Figure 2.10 Creation of a blacklist in the Shared XDMS, to be used by the PoC and Presence services

The Shared XDMS, as any other XDM server defined by OMA, is never accessed directly by XDM clients. Rather, access is managed in a secure way by the Aggregation Proxy, which is seen by other XDM servers as a reverse-HTTP Proxy.

2.8.3.2 Aggregation Proxy

In order to manage secure access to information stored in the XDM Servers, the Aggregation Proxy is the single entry point for all requests coming from and XDM Client (e.g. located at the Mobile Device). The Aggregation Proxy is responsible for contacting the right XDMS (e.g. Shared or PoC XDMS), aggregating information and providing it upon client request. All these actions must be performed in a secure way after having authenticated the requestor.

Observe that the XDM architecture is based on the exchange of XCAP documents over standard web technologies (e.g. HTTPS). As this architecture is non SIP-based, it cannot rely on IMS Authentication mechanisms, and the Aggregation Proxy must take care of user authentication.

Further details about XDM operations are provided in chapter 3.

2.8.4 OMA Presence

Together with PoC and XDM, the third OMA-defined enabler to support the PoC service is OMA Presence [7]. Presence is the third building block that completes the three main functions involved in the PoC service, as depicted in Figure 2.2. However, unlike OMA XDM, Presence is an optional feature from the PoC perspective. This is so because PoC and Presence can work independently.

Regardless, the Presence enabler provides a powerful tool to enrich the PoC service: it let users know availability of their contacts for [PoC] communications *prior* to setting up a call. This capability lets users establish PoC sessions in a non-intrusive manner, considering the Presence status of the intended callee. For example: *I will not PoC you to talk about our next date if I can see you are in a meeting at work. I would rather send you a Call-Back Request so that you can call me afterwards.*

We envisage therefore that most OMA PoC deployments will enable the Presence service to their customers, due to the enhanced user experience it delivers.

In order to provide Presence capabilities to PoC users, the client device must implement the *Watcher* and *Presence Source* functionalities respectively (with the clarifications explained in section 2.6.7), while the network operator must deploy a *Presence Service* node. We briefly present these functions in the next sections. We develop these concepts in deeper detail in chapter 4.

2.8.4.1 Presence Information

Presence information (also known as *status*) generally contains details about user's availability to communicate (*I am available for PoC communication*), but it may contain further details such as user's willingness (e.g. *Call Me / Don't Call Me*), location (*I am in Barcelona*), activities (*I am at work*) or other types of details, depending on user's preferences and client capabilities.

Presence information is stored in XML documents, which are transferred across the different entities involved in the Presence service as described below.

2.8.4.2 Presence Server

The Presence Server is responsible for conveying Presence information about users, and propagating this information to other users interested in it. Three basic mechanisms are used for this purpose:

- A client originating Presence information may *publish* a user's availability and additional Presence information at any given time to indicate whether they are able to receive PoC communications (e.g. user *online*) or not (e.g. user *offline*).
- A client may place a *subscription* in the Presence server to request Presence information updates about a contact.
- The server sends Presence information notifications towards all subscribed clients whenever the status of a particular user changes (e.g. she switches off her handset).

It is clear that the Presence service supports two types of clients: the ones *publishing* availability information (Presence Sources) and the ones *subscribing* (Watchers) to get Presence updates related to their contacts.

2.8.4.3 Presence Source

A Presence Source is a client that is capable of providing Presence information to the Presence Server. In order to implement this functionality the Presence Sources sends a SIP PUBLISH message every time Presence information changes.

2.8.4.4 Watcher

A Watcher is a Presence client that requests information to the Presence Service related to other contacts. The Watcher uses the SIP SUBSCRIBE message to indicate to the Server its interest in receiving updates regarding the status of a contact. Once the subscription is approved, the Presence Server delivers Presence information updates whenever the contact changes her status. These updates are delivered using the SIP NOTIFY transaction.

Observe that Presence information is carried as the payload of the SIP PUBLISH transaction, whenever a Presence Source sends an update towards the Presence Server, while the same information is delivered to Watchers using the SIP NOTIFY transaction.

2.8.5 Charging Entity

In general, service providers and operators are interested in charging PoC service usage. For this purpose, an external entity responsible for service charging connects to the PoC server to support charging activities. These may include checking available credit for prepaid customers or providing usage information to enable bill generation for post-paid customers.

We cover PoC charging implementation in chapter 5.

2.8.6 Device Provisioning & Management

A key element to let users access the PoC service is proper device configuration. OMA has defined a general framework to enable client configuration, provisioning and management in a standardized way. As more devices support the OMA Device Management framework and protocol [15] it is expected that this will become the common platform to deliver device management features to PoC and future mobile and multimedia services.

In particular, OMA Device Management defines a client-server protocol that enables initial configuration and update of all parameters required to access the PoC service. The operator must implement the OMA DM server endpoint to support OMA DM-enabled devices.

The OMA Client Provisioning framework [16] also supports initial handset configuration, although the full OMA DM service is required to enable modifications or application software upgrades.

In addition to standard OMA Device Management, many handset vendors and cellular operators implement legacy device configuration systems, such as sending binary-encoded Short Messages (SMS) to cellular devices in order to provision the associated application and networking settings to activate a function. These mechanisms can be used for OMA PoC configuration as well. However, it is advisable to migrate to a standard technology such as OMA DM, due to its broader applicability and scalability in the long run.

2.9 Description of OMA PoC Reference Points

2.9.1 Introduction

We have already introduced all the logical functions involved in delivering a PoC service, as described by the OMA PoC architecture [1]. In addition, [1] defines all the reference points which are used to interconnect the different functions. These reference points are required to support different types of client-server and server-server operations. In fact, every high level operation one could think of (e.g. initiating a call, creating a group, requesting the right to speak during a session, . . .) involves invoking operations over one or more PoC reference points.

For the purposes of this book, we can think an OMA reference point as a logical linkage between two OMA-defined functions. Logical functions use reference points (or *interfaces*) to request or deliver services to other entities within the PoC architecture. One or more communication protocols, such as SIP, RTP, TBCP or XCAP, may be used over a given reference point.

Table 2.2 provides the list of the eight reference points defined by OMA PoC (POC-1 to POC-8). In addition to these, there are other reference points defined by the OMA SIMPLE Presence enabler (PRS-1 to PRS-15) and by the OMA XDM enabler (XDM-1 to XDM-4) respectively. These are described in the next chapters. A brief description of the functionality supported by each reference point is provided below.

Figure 2.11 provides an at-a-glance mapping between PoC reference points and the corresponding protocol used to implement the reference point's functionality. Observe that, for sake of completeness, OMA Presence and XDM reference points shown here also display the associated protocol.

Table 2.2 OMA PoC reference points

Reference Point Name	Usage	Protocol(s)
POC-1	PoC Client to SIP/IP Core Session signalling	SIP
POC-2	SIP/IP Core to PoC Server Session signalling	SIP
POC-3	Media, Talk Burst Control and Media Burst Control	RTP, RTCP, TBCP
POC-4	Media, Talk Burst Control and Media Burst Control between networks	RTP, RTCP, TBCP
POC-5	Shared XDMS to PoC Server	XCAP
POC-6	SIP/IP Core to PoC XDMS	SIP
POC-7	PoC XDMS to Aggregation proxy	XCAP
POC-8	PoC Server to PoC XDMS	XCAP
IP-1	SIP/IP Core to SIP/IP Core	SIP

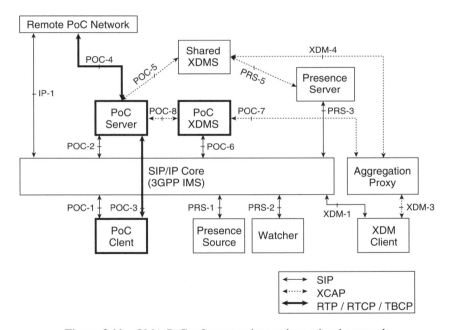

Figure 2.11   OMA PoC reference points and associated protocols

2.9.2 Reference Point POC-1: PoC Client – SIP/IP Core

This reference point is used by the client to exchange SIP signalling with the SIP/IP Core. The features supported by the POC-1 interface are:

- User authentication, authorization and registration with the SIP/IP Core
- Notification of PoC capabilities during registration
- SIP signalling compression

- Service and user reachability and SIP signalling routing
- Relay of PoC service service settings published by the client towards the PoC Server
- Supports exchange of all SIP signalling required for PoC service operation: session initiation (both originating and terminating cases), sending/receiving Instant Personal Alerts, sending/receiving Group Advertisements, subscriptions and notifications of PoC events . . .
- Provides integrity protection and -optionally- confidentiality protection of SIP signalling.

2.9.3 Reference Point POC-2: SIP/IP Core – PoC Server

The POC-2 reference point connects the PoC Server with the SIP/IP Core. This interface is used to deliver SIP signalling sent by the PoC Client to the PoC Server (and vice-versa), after proper authentication, authorization, service discovery and routing tasks are performed by the SIP/IP Core. When the PoC Service is deployed over a 3GPP compliant cellular network, the SIP/IP Core function is performed by the IP Multimedia Subsystem (IMS), and the OMA POC-2 reference point maps to the well-known 3GPP ISC (IMS Service Control) interface defined between the IMS and any SIP Application Server.

The following operations are supported by the POC-2 reference point:

- Delivery of Third-Party Registration information from the SIP/IP Core towards the PoC Server, including notification of PoC capabilities supported by the client,
- Provides charging information to the PoC Server,
- Address resolution services for requests containing TEL URI addresses in their SIP Request URI[6],
- Delivery of Presence information from the PoC Server to the Presence Server (and vice-versa) if this function is supported (refer to section 2.6.7),
- Subscriptions and notification of changes of documents stored in the PoC XDMS or the Shared XDMS, if this function is supported by the service (refer to section 2.6.6),
- Relay of PoC service service settings published by the client towards the PoC Server,
- Relay of PoC service signalling between the PoC Client and the PoC Server (and vice versa).

2.9.4 Reference Point POC-3: PoC Client – PoC Server

Reference point POC-3 connects the PoC Client and the PoC Server. This interface is used for all media plane communications. The three functions implemented over this interface are:

---

[6] Routing of SIP messages within a SIP network is performed based on SIP URI's (e.g. sip:joe.doe@wirelessfuture. com). A PoC endpoint may dial a plain telephone number using the TEL URL format [18] (e.g. tel:+358 12 34 56) to address the callee of a PoC session. In such case, the SIP/IP Core performs an automatic translation into the corresponding SIP URI address assigned to that client, in order to properly route the request.

- Media transport. Delivery of digitized audio over RTP between the PoC Client and the PoC Server,
- Floor control. Clients and servers use interface POC-3 to request and authorize permission to speak during a PoC session respectively. TBCP is used for this purpose.
- Quality feedback. Typically, RTP endpoints (i.e. PoC Clients and Servers) implement the Real-time Control Protocol (RTCP) to inform the remote endpoint about the quality of media delivery (e.g. delivered packets, lost packets, estimated bandwidth).

Depending on user plane routing configuration, the POC-3 reference point connects the PoC Client with the Controlling or the Participating PoC function. In some cases, direct media flow between the PoC Client and the Controlling PoC function (when they are located in different domains) may not be allowed due to operator policies. Therefore, the POC-3 interface must be supported between the PoC Client and the P-PoC function, and it may be implemented between the PoC Client and the C-PoC function if interoperator agreements allow.

2.9.5 Reference Point POC-4: PoC Server – PoC Server

The POC-4 reference point is used to provide media plane and floor control communication between the Participating PoC function and the Controlling PoC function, when a PoC session involves two or more PoC servers. RTP, RTCP and TBCP are the protocols used in the POC-4 reference point.

2.9.6 Reference Point POC-5: PoC Server – Shared XDMS

The POC-5 reference point connects the PoC Server with the Shared XDMS. XCAP is the protocol used in POC-5. This interface is used by the PoC Server to retrieve lists of URI's stored in the Shared XDMS. As an example, when a PoC group XML document contains a link to an external URI list XML document stored in the Shared XDMS, the PoC Server uses this reference point to retrieve that URI list.

2.9.7 Reference Point POC-6: SIP/IP Core – PoC XDMS

This interface is used to enable subscription to changes in XML documents. This feature can be used by a PoC Server to receive timely updates of XML documents.

The reference protocol used in the POC-6 reference point is SIP. The PoC Server uses the SIP SUBSCRIBE transaction to activate a subscription to changes in XML documents. Changes are notified using the SIP NOTIFY transaction.

Subscription and notification of changes in XML documents is a PoC Server optional feature.

NOTE: The ability to subscribe to changes in XML documents is supported in PoCv1.0.2. However, this feature was finally dropped for Presence v1.1 and XDMv1.1, as it is based on an Internet-Draft which was eventually discontinued. Hence, it is not clear whether this feature will stay or not, in case new revisions of the PoCv1 standard take place in the future.

In PoCv2, this feature will be based on a new general mechanism that is currently under definition at IETF. The reader is referred to chapter 3 for further details.

2.9.8 Reference Point POC-7: Aggregation Proxy – PoC XDMS

The POC-7 reference point connects the Aggregation Proxy and the PoC XDMS. XCAP is the protocol used between both endpoints. This interface is used by the Aggregation Proxy to perform operations (creation, retrieval, modification, deletion) in XML documents stored in the PoC XDMS upon user request (the Aggregation Proxy centralizes all XCAP operations initiated from an XDM Client).

2.9.9 Reference Point POC-8: PoC Server – PoC XDMS

The POC-8 reference point is used by the PoC Server to retrieve documents stored in the PoC XDMS. The PoC Server uses XCAP to perform POC-8 operations. The PoC server uses reference point POC-8 whenever it requires retrieving an XML document stored in the PoC XDMS. As an example, when a user initiates a Pre-Arranged PoC group session, the Controlling PoC server will download the PoC group XML document stored in the PoC XDMS to retrieve all required information required to proceed with session initiation.

Observe that since the PoC Server is located in a trusted domain under the operator's control, there is no need to route PoC Server requests through the Aggregation Proxy. Rather, a direct interface between the PoC Server and the PoC XDMS is implemented for scalability and performance reasons (this is the same case for reference point POC-5).

2.9.10 IP-1 Reference Point: Interconnecting SIP (PoC) Networks

The IP-1 interface is defined between SIP/IP Core functions provided by different network domains. This interface is not defined by OMA PoC, as it relies on the functionality provided by the underlying SIP / IP Core function. Whenever a PoC session involves communication between two or more SIP domains, this logical function is required. The typical use case requiring IP-1 functionality is a PoC session (1-to-1 or 1-to-Many) where there are at least two subscribers of different network operators, as shown in Figure 2.12.

Observe that IP-1 functionality is managed directly by the SIP/IP Core. Therefore, its definition is out of the scope of OMA activities. When PoC is deployed on top of a 3GPP IMS platform, the IP-1 capability is supported by the 3GPP Mm reference point defined in [19]. In Figure 2.12, the flow of the initial SIP INVITE message will follow the path indicated below (assuming that Joe initiates a PoC session with Mary):

1. PoC Client A → SIP / IP Core A
2. SIP / IP Core A → PoC Server A
3. PoC Server A → SIP / IP Core A (the PoC Server determines that the served user is in a remote network)
4. SIP / IP Core A → SIP / IP Core B (the SIP / IP Core A determines that 'domainB.com' domain is in a remote network and uses SIP/IMS routing capabilities to deliver the SIP INVITE message to the domain of the called user)

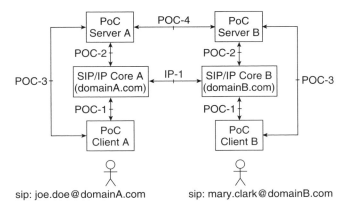

Figure 2.12   IP-1 Reference Point

5. SIP / IP Core B → PoC Server B (as a PoC message, SIP/IP Core B will deliver it to PoC Server B)
6. PoC Server B → SIP / IP Core B
7. SIP / IP Core B → PoC Client B.

At this stage, PoC Client B may issue a temporary or final answer, which will be proxied backwards following exactly the same route (in the opposite direction).

2.9.11 XDM Reference Points

The OMA XDM enabler defines four specific reference points, namely: XDM-1 to XDM-4. Among them, XDM-3 and XDM-4 are the most important ones from the PoC service perspective.

- XDM-1 is defined between the XDMC and the SIP/IP Core, to support subscription to changes in XML documents. The XDM enabler will finally not support this feature, so this interface is not used in XDMv1.1 [20]. The interface numbering is kept for backwards compatibility with the architecture defined in XDVv1.0 and XDMv1.0.1.
- XDM-2 was initially defined to support subscription to changes in documents stored in the Shared XDMS. Hence, it will also be withdrawn in the next expected revision of OMA XDM (XDMv1.1 [21]).
- XDM-3 is one of the key XDM interfaces. It connects the XDMC with the Aggregation Proxy, and it is used by the XDMC to perform operations in XML documents stored in any XDMS. Therefore, XDM-3 is used to create, modify or delete PoC groups, PoC user access policies stored in the PoC XDMS or URI lists stored in the Shared XDMS. XDM-3 is also used to modify XML documents stored in the Presence XDMS and RLS XDMS, which are XDM nodes specific to the Presence service (they are described in further detail in chapter 4). All XDM-3 operations are performed via the Aggregation Proxy, which is in charge of authenticating the user before authorizing any XCAP operation.

- XDM-4 reference point is used by the Authentication Proxy to perform XCAP operations in XML documents stored in the Shared XDMS on behalf of an authorized XDM user. As such, it offers the same type of functionality as the POC-7 interface.

2.9.12 Presence Reference Points

In this section we will provide a short description of the Presence interfaces depicted in figure 2.7. Additional details about all OMA Presence reference points are provided in chapter 4.

- PRS-1 is used by the Presence Source to publish Presence information about the end user.
- PRS-2 is used by the Watcher to subscribe and receive Presence notifications about other users.
- Observe that the Presence Source and Watcher functions may reside at the mobile handset that stores the PoC Client and XDMC functionality. Note that implementing Watcher capabilities does not require implementing Presence Source capabilities. Hence, a PoC device may display Presence information about PoC contacts without publishing Presence information about the local PoC user (e.g. the PoC Server may publish PoC status on behalf of the user as described in section 2.6.7 and chapter 4).
- PRS-3 connects the Presence Server to the SIP/IP Core. It delivers Presence publications and subscriptions to the Presence Server, and it collects Presence notifications to be delivered to Watchers.
- PRS-5 is used by the Presence Server to retrieve URI lists stored in the Shared XDMS. This interface provides the same type of functionality as POC-5. URI lists are used in the Presence service to define Presence authorization policies. A Presence authorization policy document defines which watchers are authorized to receive Presence information about a Presentity.

2.10 Summary and Conclusions

In this chapter we have presented the end-to-end OMA PoC architecture. We have presented PoC as a SIP-based service deployed on top of a SIP / IP Core function that provides authentication, authorization and routing capabilities. The two basic elements of the PoC architecture are the PoC client and the PoC server. These elements support basic session setup, 1-to-1 and 1-to-many half-duplex communication features, and all PoC capabilities in general.

In order to support group and policy management operations, the XDM architecture is required, including at least the XDMC, the Aggregation Proxy and the PoC XDMS. The Shared XDMS provides centralized URI list management.

The third key enabler supporting the PoC service is OMA Presence. Presence enriches the end-user experience by providing dynamic information about a user's contacts availability and willingness to communicate.

The four key protocols supporting PoC, Presence and XDM operations are SIP (supports service-related signalling), RTP/RTCP (supports media delivery), TBCP (for floor control operations) and XCAP (to manage of XML documents).

We have described all the interfaces, functions and relationships between all the PoC, XDM and Presence nodes required to deliver an end-to-end PoC service.

In the next chapters we will provide a deeper degree of detail in topics such as OMA XDM and OMA Presence enablers (chapters 3 and 4 respectively), deployment of OMA PoC over a 3GPP IMS platform supporting the SIP / IP Core function (chapter 5) and detailed PoC operations and signalling flows (chapter 6).

2.11 References

[1] OMA Push-to-Talk over Cellular (PoCv1.0.2): 'PoC Architecture'; September 2007.

[2] J. Rosenberg, Schulzrinne, H., Camarillo, G., Johnston, A., Peterson, J., Sparks, R., Handley, M. and E. Schooler: 'SIP: Session Initiation Protocol'. RFC 3261. IETF, June 2002.

[3] H. Schulzrinne, S. Casner, R. Frederick, V. Jacobson: 'RTP: A Transport Protocol for Real-Time Applications'. RFC 3550. IETF, July 2003.

[4] OMA Push-to-Talk over Cellular (PoCv1.0.2): 'PoC User Plane'; September 2007.

[5] J. Rosenberg: 'The Extensible Markup Language (XML) Configuration Access Protocol (XCAP)'; RFC 4825, IETF, May 2007.

[6] OMA XML Document Management (XDMv1.0.1) 'XML Document Management Architecture'; November 2006.

[7] OMA SIMPLE Presence (Presence v1.0.1) 'Presence SIMPLE Architecture'; November 2006.

[8] OMA Push-to-Talk over Cellular (PoCv1.0.2): 'PoC Control Plane'; September 2007.

[9] A. Niemi, 'Session Initiation Protocol (SIP) Extension for Event State Publication', RFC 3903, October 2004.

[10] B. Campbell, J. Rosenberg, H. Schulzrinne, et al., 'Session Initiation Protocol (SIP) Extension for Instant Messaging', RFC 3428, December 2002.

[11] OMA Device Management (DM): 'Enabler Release Definition for OMA Device Management'; Candidate Version (v1.2), February 2006.

[12] OMA 'Dictionary for OMA Specifications (version 2.4)' (Approved Version); July 2006.

[13] G. Camarillo, J. Ott, K. Drage, 'The Binary Floor Control Protocol (BFCP)', Internet-Draft (work in progress), November 2005.

[14] OMA 'Utilization of IMS capabilities – Requirements'; August 2005.

[15] OMA 'Utilization of IMS capabilities – Architecture'; August 2005.

[16] OMA Device Management (DMv1.2) 'Device Management Protocol'; June 2005.

[17] OMA Client Provisioning (CPv1.1) 'Client Provisioning Architecture'; April 2005.

[18] H. Schulzrinne: 'The tel URI for Telephone Numbers', RFC 3966, December 2004.

[19] OMA: 'Utilization of IMS Capabilities – Requirements (version 1.0)' (Approved Version); August 2005.

[20] OMA Push-to-Talk over Cellular (PoCv1.0.2): 'Push-to-Talk over Cellular Requirements Document'; June 2006.

[21] OMA XML Document Management (XDMv1.1) 'XML Document Management Architecture'; August 2007 (Work In Progress).

3

The OMA XML Document Management (XDM) Enabler

3.1 Introduction

OMA started activities around PoC standardization by creating the PoC Working Group in mid-2003. Later on in 2003 the *Presence and Availability Group* (PAG WG) was created to leverage IETF and 3GPP/3GPP2 work in the Presence area, and define a new mobile standard to exchange dynamic information such as user status, location and capabilities.

Shortly after that, OMA realized that both WGs needed to develop some sort of support to manage groups, or lists of users and contacts. Instead of developing two group management functionalities, it was clear that both PoC and Presence could benefit from a common approach to group management.

With these ideas in mind, in late 2003/early 2004 OMA created the Group Management *Work Item* to define a common *Group Management Enabler* [1]. This should provide a common framework to support several (present and future) applications, letting each service enabler extend the core functionality. In order to ensure that all PoC and Presence requirements would be taken into account, and that the required timelines would be met by both groups, it was decided that Group Management standardization would happen within the same PAG WG that was taking care of the Presence enabler.

During the standardization work, it was realized that the approach to group management could be generalized still further: the group management concept was developed around sharing of XML documents which may be used not only to define groups, but also policies and service settings. Hence, in mid-2004 OMA renamed initial Group Management work into the more generic 'XML Document Management (XDM) Enabler'. The overall XDM architecture, therefore, provides not only Group Management but also Policy and Access Control support to other services such as Presence and PoC. In addition to the more general approach, a significant simplification of the scope of XDM activities was also carried out (e.g. Messaging capabilities were de-scoped for the first version of the enabler), in order to ensure timely availability of a comprehensive specification.

After the completion of standardization activities, the XDM enabler lets PoC users create, modify, retrieve and delete XML documents and their contents. Different XML document types

Multimedia Group Communication. Andrew Rebeiro-Hargrave and David Viamonte Solé
© 2008 John Wiley & Sons, Ltd.

are used to support different application-specific purposes: for example, definition of PoC groups, PoC policies (e.g. *who is allowed to call me?*) or Presence lists (e.g. *which contacts am I interested in getting Presence information from?*). Theoretically, given XML extensibility features, the OMA XDM enabler may support any type of application-specific configuration that can be expressed via a well-formed, valid and semantically meaningful XML document.

The core of the communications between elements of the XDM architecture is supported by IETF-defined *XML Capability Access Protocol* (XCAP) [2]: OMA XDM is an XCAP compliant architecture. Several OMA extensions have been defined to complement basic XCAP so that it fulfils some specific requirements.

It is worth understanding what an XCAP document looks like and what type of operations can be performed using XCAP/XDM operations. Consider the following sample document that describes a PoC group:

```
<?xml version="1.0" encoding="UTF-8"?>
<group xmlns="urn:oma:xml:poc:list-service"
       xmlns:rl="urn:ietf:params:xml:ns:resource-lists"
       xmlns:cr="urn:ietf:params:xml:ns:common-policy"
       xmlns:xsi="http://www.w3.org/2001/XMLSchema-
       instance">
  <list-service uri="sip:my_colleagues@wirelessfuture.com">
    <display-name xml:lang="en-us">My Colleagues</display-name>
    <list>
     <entry uri="tel:+43012345678"/>
     <entry uri="sip:mary@wirelessfuture.com"/>
    </list>
    <cr:ruleset>
     <cr:rule id="a7c">
       <cr:conditions>
         <is-list-member/>
       </cr:conditions>
       <cr:actions>
         <allow-anonymity>true</allow-anonymity>
       </cr:actions>
     </cr:rule>
    </cr:ruleset>
  </list-service>
</group>
```

This document describes a PoC group (e.g. owned by Joe – sip:joe.doe@wirelessfuture.com, although this information is not maintained in the document's content). The group has a group name (sip:my_colleagues@wirelessfuture.com). Joe has included two members in the group: for the first member, Joe knows her telephone number (+43012345678), while for the second contact, Joe has used her associated SIP URI (sip:mary@wirelessfuture.com).

With OMA XDM, Joe may perform the following operations:

- Modify the *display name* ('My Colleagues') associated to the group
- Add or remove members from the group members list
- Define a new set of policies associated to the group (e.g. deny anonymous access to group sessions)
- . . .

In this chapter, we will describe how all these and other operations are performed within the XDM framework, and how OMA XDM specifies a well-known and hierarchical structure, so that document owners (such as Joe) can keep track of all their XML documents describing application settings and policies.

3.2 The OMA XDM Architecture

3.2.1 Introduction

The main purpose of XDM is to define the architecture of servers storing XML documents, and mechanisms to communicate information contained in those XML documents to client and/or server entities. XML documents stored in XDM Servers contain information that is required by applications in order to determine how to process and serve user requests. These XML documents may contain, for example, access control lists, PoC groups or Presence subscription lists.

Figure 3.1 shows the OMA XDM Architecture [3]. Three main elements build up the *core* of the system: the *Shared XDM Server* (XDMS), the *XDM Client* (XDMC) and the *Aggrega-*

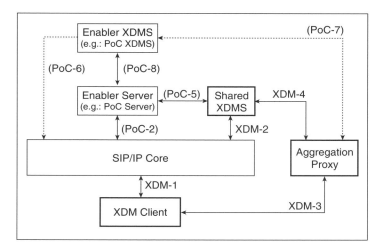

Figure 3.1   The OMA XDM Architecture[1]

---

[1] NOTE: At the time of writing (September 2007) OMA is about to approve the second revision of the XDMv1 enabler (OMA XDMv1.1). An important architectural change from the depicted picture is the removal of interfaces XDM-1 and XDM-2. Although interface numbering will be kept to ensure backwards compatibility with XDMv1.0 and v1.0.1 naming conventions, its functionality will finally not be included in the first version of the OMA XDM enabler, as described in section 3.2.2.2.

tion Proxy. Although the detailed behavior of these elements will be presented below, we provide an at-a-glace view of the goal of these three main logical functions defined by OMA XDM:

- The Shared XDMS stores XML documents which can be shared (reused) by several different applications (e.g. PoC, Presence, Messaging).
- The XDMC manages XML documents stored in the Shared XDMS (or in any other XDM Server) so that the end user is able to create, retrieve, modify or delete application-specific settings stored in those XML documents.
- The Aggregation Proxy is in charge of authenticating the XCAP user, to ensure that communication between the XDMC and the XDM Servers is handled in a secure way. In some cases, information may be disseminated across different XDM Servers. In that event, the Aggregation Proxy collects and aggregates the information prior to its delivery to the XDMC (hence, the name of this node).

The goal of the XDM architecture is to provide a framework to let users manage XML documents that define service-specific settings and policies. Thus, the XDM enabler only makes complete sense when it is connected to 'real' SIP-based services such as PoC, Presence or SIP/SIMPLE Messaging.

While documents stored in the Shared XDMS can be shared across different SIP applications, in most cases each application uses a different type of XML document (and schema) to define application-specific policy or configuration. As an example, the XML document used to define a PoC group is different from the XML document used to describe a Presence policy document. Thus, in general, each SIP service incorporates an application-specific XCAP/XDM server, which manages application-specific XML documents. Figure 3.1 shows how this architecture works using OMA PoC as a sample SIP application. In particular, OMA defined service enablers (e.g. PoC, Presence, SIMPLE Messaging) incorporate the description of (at least) one enabler-specific XDMS function. For example, the PoC-specific XDMS is the PoC XDMS.

OMA XDM defines the three main XDM logical functions presented above, plus the 'core' XDM interfaces that connect these functions. These interfaces (XDM-1, XDM-2, XDM-3, XDM-4) are specified in [3, 4], and we present them in section 3.3.

In addition to 'core' XDM interfaces, there is the need to connect external SIP applications with the XDM functions: an Application Server (e.g. a PoC server) should be able to retrieve XML documents stored in the Shared XDMS and/or in the enabler-specific XDMS. For this purpose, additional reference points must be defined to connect each SIP service with common OMA XDM logical functions. In order to ensure a future-proof architecture, reference points connecting specific SIP applications with general XDM functions are defined under the external OMA enabler specification. For example, POC-5, POC-6, POC-7 and POC-8 reference points are defined by the OMA PoC specification [5], rather than by the OMA XDM enabler.

It is worth observing that all XDM transactions consist basically on management (creation, retrieval, modification, deletion) of XML documents stored in a XDM server, using XCAP mechanisms.

XDM architecture elements and interfaces are described in further detail in the following sections.

3.2.2 XDM Client (XDMC)

End user devices that support a SIP application (e.g. a PoC client) generally implement the XDM Client function. The XDMC lets end users manage XML documents stored in the network and, this way, configure their application settings. Users do not need to care about complex operations in the background, since the XDMC user interface should map simple user operations (e.g. *I want to add a contact from my 'buddy list' to my 'chess team' PoC group*) into the required XCAP or SIP transactions.

3.2.2.1 XCAP Document Management Operations

The XDMC must implement the XCAP protocol to let the user create, retrieve, modify and delete XML documents stored in any XDM Server (e.g. Shared XDMS, PoC XDMS). These operations can be performed not only at a 'document' level (e.g. upload or delete a complete XML document), but also at the 'node' level. A 'node' is an XML resource within an XML document (e.g. an XML element or an attribute of an XML element), which can be managed via XCAP operations. This level of granularity lets the XDMC modify an XML document without having to download and upload a whole copy of it whenever a minor operation has to be performed (for example: adding a new contact to an existing PoC group owned by the user).

The XDMC does not have direct connectivity to XDMS's. Rather, all these operations are proxied through the Aggregation Proxy. Since the Aggregation Proxy is in charge of authenticating the XDMC user, the XDMC must support the authentication mechanisms (e.g. HTTP Digest) defined in [4]. The XDMC may have access to USIM/ISIM information in order to retrieve authentication data and credentials. XCAP identity and XCAP authentication is explained in detail in 3.5.

When the XDMC is implemented in a cellular device (e.g. a PoC handset) communication between the XDMC and the Aggregation Proxy takes place over a potentially narrowband, error-prone and slow wireless link. The XDMC may support HTTP compression in order to improve efficiency of XCAP operations in such case [4].

3.2.2.2 Subscription to Changes in XML Documents (Withdrawn)

The OMA XDM enabler initially defined a mechanism to let XDM clients subscribe to changes in XML documents stored in any XDM server. The idea was to enable real-time updates to subscribing clients or servers, of modifications in any XML document. As an example, a PoC server hosting a Pre-Arranged PoC Group session may subscribe to changes in the PoC group document. This way, if the group owner removes an entry from the group member list, the PoC server could remove the recently deleted member from the session.

This feature, however, was based on an IETF Internet Draft that was eventually dropped. Hence, OMA will remove this feature from the second revision of the OMA XDMv1 enabler (OMA XDMv1.1) [24], and from PoC and Presence enablers where this feature was also included.

IETF and OMA have already initiated activities to define a more generic mechanism to subscribe to changes in XML documents [25]. It is expected that such mechanism will be incorporated into version 2 of OMA XDM, PoC and Presence enablers which are, at present, at an early standardization and interoperability stage.

3.2.2.3 XDMC Implementation

The XDMC is a 'logical' function that may be implemented in different ways. In general, cellular handsets supporting PoC communications will include an XDMC module, to let end users manage PoC groups and access policies (e.g. PoC *black lists*).

In addition to the XDMC residing in a mobile device, there are other XDMC implementations that are of interest:

- Operators may offer web-based access to XCAP operations. With this configuration, end users can configure their SIP application settings using a standard web browser. The advantage of this alternative is that users can comfortably access and manage their PoC groups and policies from their home or professional computer (e.g. laptop PC or desktop computer), and enjoy the improved processing power and rendering capabilities when compared to their regular PoC handset. Given that XCAP is an HTTP-based mechanism, it should be relatively easy to provide this feature to end users.
- Additionally, an Application Server can implement the XDMC functionality (server-based XDMC). For the time being, however, we do not expect Application Servers massively implementing the XDMC function, since its application usages and benefits remain a bit unclear at the time of writing.

Figure 3.2 povides a general overview of the main features that a client-based XDMC should support. Observe that apart from technical features such as access to authentication information stored in the ISIM or XCAP communications, the XDMC has the key role of offering an appropriate *Graphical User Interface* (GUI) to the end user, to let him easily and naturally manage groups, policies and XML documents stored in the XDM environment.

3.2.3 Aggregation Proxy

The Aggregation Proxy is the entry point for the XDMC into the XDM service. Its main purposes are:

- Support XCAP user authentication when accessing XML documents stored in the XDM environment. Observe that, since XDM-3 interface does not use the SIP/IP Core infrastructure (e.g. IMS), the Aggregation Proxy must use its own mechanism to authenticate the XDM user.
- Provide XCAP message routing between the XDM client and the proper XDM server.
- Support aggregation of information from the XDM servers to the XDMC.

The Aggregation Proxy may support HTTP compression [4] over the XDM-3 interface. This mechanism is only used when the XDMC supports it as well.

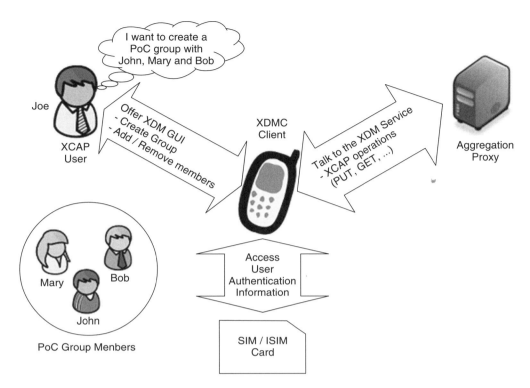

Figure 3.2   XDMC logical functions

In fact, the XDMC sees the Aggregation Proxy as a single entity managing all XML documents. For each XDMC request, the Aggregation Proxy must determine to which XDMS the request must be routed. As an example, if an XDMC requests access to a URI List document, the Aggregation Proxy should redirect this operation to the Shared XDMS. In contrast, when the Aggregation Proxy detects that the XDMC wishes to perform an XCAP operation over a PoC group XML document, it redirects the request to the PoC XDMS.

Some XDMC requests may require communication with several XDMSs within the same XCAP transaction (refer to section 3.6.2 for further details). In this case, the Aggregation Proxy is seen as an HTTP Reverse Proxy from the XDMC point of view. It must fork the requests to the required XDMSs and deliver an aggregated response to the XDMC. Because of this capability, we can think that – from the XDMC perspective – the Aggregation Proxy behaves as a unique XDMS supporting all XCAP applications.

The OMA XDM Architecture [3] indicates that the Aggregation Proxy may support charging of XDM services. However, OMA does not define any capability to support such functionality. Additionally, since XDM is an enabling technology that supports other services such as PoC and Presence, it seems more appropriate to place charging functions in those services[2].

---

[2]PoC and Presence service charging may be complemented by incorporating some policy functions into the XDM service, to avoid malicious usage. As an example, the operator may wish to prevent users from performing XCAP operations at an extremely high frequency. A user changing XML documents continuously would cause signalling overhead (XCAP transactions plus SIP notifications to users subscribed to changes in XML documents).

3.2.4 Shared XDMS and Basic Introduction to URI Lists

The main purpose of the Shared XDMS is to store *URI Lists*. A URI List is an XML document that contains a list of SIP URIs, TEL URIs (TELephone Uniform Resource Identifier) or a mix of SIP URIs and TEL URIs[3]. URI Lists are a convenient way for a user to group together a set of URIs in a common location. Lists stored in the Shared XDMS may be 'service-agnostic': the list can be reused by several services to build-up application-specific groups that have a complete meaning within the scope of that service. The following example shows a sample URI List:

```
<list name="close-friends">
  <display-name>Close Friends</display-name>
  <entry uri="sip:mary.watson@wirelessfuture.com">
    <display-name>Mary</display-name>
  </entry>
  <entry uri="ron@wirelssfuture.com"/>
    <display-name>Ron</display-name>
  </entry>
  <entry uri="tel:5678;phone-context=+358012349999"/>
</list>
```

The above list contains three `<entry>` XML elements: two SIP URIs and one TEL URI. Some elements contain the optional `<display-name>` element. The list is called 'close-friends', and includes a list-wide *display name* as well. Such a URI list would be stored in a XML document located in the Shared XDMS. The XML document provides all namespaces declarations, and it may include zero or more `<list>` elements. Such document is called a 'resource-lists' document [7], and is described in further detail in section 3.6.3.1.

We will use the term *URI List* and *Resource List* [7] interchangeably. A *Resource List* document is a document that contains a list of *resources*. Resources are addressed via URIs; therefore, a resource list generally contains zero or more URI Lists. Each entry in a URI List points to an addressable *resource*: a contact's PoC address, a service, an IVR system, a web page, or even another list – to build-up nested lists. As mentioned above, within the scope of PoC, Presence and XDM OMA enablers, URI Lists contain SIP URIs and/or TEL URIs.

The benefit of storing URI Lists in the Shared XDMS is that lists become service-agnostic. Thus, a common list can be reused across several applications, and once a change is made to the shared list, this change is automatically propagated to all services, without the user having to modify N application-specific lists stored in N application-specific XDMSs.

Apart from the mandatory URI List support, the Shared XDMS may support storage of *Group Usage List* documents. The Group Usages List document maintains a mapping of the

---

[3] In fact, a URI List may contain other URI schemas (e.g. http, mailto). However, in the framework of OMA PoC and XDM only SIP or TEL URI's are meaningful for the service.

list of user-owned PoC groups and its associated behavior (i.e. Chat – dialed-in – vs. Pre-Arranged – dialed-out – PoC Groups).

The Shared XDMS supports creation, retrieval, modification and deletion of XML documents (URI Lists or Group Usage Lists), elements and/or attributes within an XML document upon XDMC request. Usage of Resource Lists (URI Lists) and Group Usage Lists documents stored in the Shared XDMS is explained in further detail in section 3.6.3.

3.2.5 External Enabler-specific Entities involved in the XDM Architecture

The last two elements involved in the XDM architecture are the external (not defined by OMA XDM) 'enabler-specific XDMS' and the 'enabler server'.

3.2.5.1 'Enabler-specific' XDMS

As mentioned, each SIP service defined by OMA will generally require the definition of one or more *XCAP applications*, to support users managing their service data and settings. For example, the PoC service defines two new XCAP applications: *PoC groups* and *PoC user access policy*.

In order to support application-specific XML documents, the OMA XDM architecture leaves open for each OMA enabler the definition of an external XDM server. With this approach, documents stored in the Shared XDMS are kept general, and the architecture remains future-proof: new XCAP applications and Enabler-specific XDMS's can be defined in the future, without the need to update the Shared XDMS every time a new SIP application is created. Still, application-specific XML documents may contain links to documents stored in the Shared XDMS if needed.

As a consequence of this approach, XDM enabler-specific functions are always defined by the corresponding enabler-specific OMA specification documents [3, 5, 9]. We have already explained that the PoC XDMS supports the XDM functions required to let users manage the PoC service. The OMA Presence enabler defines two additional XDMS functions [9, 10, 11]:

- The Presence XDMS stores Presence policies
- The RLS XDMS stores Presence subscription lists.

The existence of two enabler-specific XDMS's in OMA Presence is because OMA specifies two well-defined Presence service entities: the Presence server and the Resource List server (RLS). From the SIP point of view, each node implements a different logical function. Thus, two different XDMS's are required.

When we talk about enabler-specific XDMS's, we refer to *logical* functions. It is perfectly feasible that two or more functions are implemented in the same physical node. In particular, the XCAP framework defines a structured hierarchical approach to how XML documents should be stored in a XDM/XCAP server. Thus, it is feasible that a single server manages XML documents pertaining to different applications, without confusing or mixing documents from different services or users. As an example, the Shared, PoC, Presence and RLS XDMS functions might be implemented in the same machine.

Observe that, regardless of the preferred approach — collapsed 'all-in-one' XDMS vs. distributed XDMSs — most of the core functionality of all XDMS's (Shared, PoC, Presence, RLS) is very similar: a server capable of processing management of XML documents over HTTP. Each XDMS must be able to perform the following actions:

1. Authorize user access to XML documents stored in the XDMS
2. Validate XML documents stored in the XDMS (gramatically, syntactically and semantically valid documents)
3. Support all XCAP applications required to implement the specific XDMS function defined by OMA.

The main difference between two XDMS's resides in the third bullet: a PoC server supports *PoC Groups* and *PoC User Access Policy* XCAP applications, while a Presence XDMS supports *Presence Authorization Rules* documents. However, the first two capabilities are similar for any XDMS (and even for any XCAP server).

Observe that, within the OMA XDM framework, the fact that several XDMSs are collapsed or distributed in a given implementation is totally transparent to the XDMC. Effectively, from the XDMC point of view there is one single XDMS. This is so because the XDMC uses a single connection against the Aggregation Proxy. The Aggregation Proxy uses the HTTP URI information contained in XDMC's requests to decide which should be the right XDMS responsible for processing a single transaction.

3.2.5.2 Enabler Server

In addition to the specification of the enabler-specific XDMS, OMA defines a set of reference points that connect an enabler server (e.g. the PoC server) with the XDM architecture. Since XML documents stored in a XDMS (be it the Shared XDMS or an enabler XDMS) define service policies and configurations, it seems logical that the application server (e.g. the PoC Server) should be able to retrieve those XML documents in order to apply the policies and settings that they define.

In summary, the list below provides an overview of all the capabilities that must be implemented with the collaboration of XDM and enabler-specific nodes:

• There is an XCAP-based interface between the enabler-specific server and the Shared XDMS. This interface is used to retrieve shared XML documents (e.g. URI lists) stored in the Shared XDMS,
• There is an XCAP-based interface between the enabler-specific server and the enabler-specific XDMS. This interface is used to retrieve enabler-specific XML documents,

The enabler server must be able to communicate with the enabler XDMS and the Shared XDMS in order to retrieve XML documents. These documents are necessary for the SIP service to access per-user information in the process of servicing a user's request.

- There is an XCAP-based interface defined between the enabler-specific XDMS and the Aggregation Proxy to let the XDMC manage XML documents (e.g. URI lists) stored in the enabler-specific XDMS.

3.2.5.3 SIP/IP Core

The OMA XDM enabler assumes that a generic SIP/IP Core function is used. This function provides SIP user authentication, signalling compression, message routing and service discovery. OMA XDM assumes that this function is provided by an external platform such as 3GPP IMS. For additional information about implementation of OMA enablers over a 3GPP IMS the reader is referred to chapter 5.

3.3 XDM Reference Points

3.3.1 Introduction

The XDM Architecture [3] defines four reference points. Table 3.1 summarizes the endpoints connected by each reference point, and the protocol used between both endpoints.

In the following sections we provide a description of the usage of each OMA XDM reference point. PoC and Presence reference points are described in section 3.3.5.

3.3.2 Reference Points XDM-1 and XDM-2 (Withdrawn)

As described in section 3.2.2.2, subscription to changes in XML documents is a feature to be dropped in XDMv1.1 revision. This was the only feature supported by XDM-1 and XDM-2 reference points, which are hence not implemented.

3.3.3 Reference Point XDM-3: XDM Client – Aggregation Proxy

This is the entry point to XDM services for the XDM client. The required functions supported by the XDM-3 reference point are:

Table 3.1 OMA XDM Reference Points

XDM Reference Points		
Reference Point	Endpoints	Protocol
XDM-1	XDM Client – SIP/IP Core	SIP
XDM-2	SIP/IP Core – Shared XDMS	SIP
XDM-3	XDM Client – Aggregation Proxy	XCAP
XDM-4	Aggregation Proxy – Shared XDMS	XCAP

- Mutual authentication of the XDM client and the Authentication Proxy
- Management of XML documents (creation, retrieval, modification, deletion of documents and/or its contents)
- Compression of HTTP messages exchanged over the interface [4]. This feature is optional, most relevant when the XDMC connects to the XDM service over a cellular – potentially narrowband and error-prone – environment. When implemented, HTTP compression is done according to [12].

Communication integrity and protection of confidentiality are performed by implementing HTTP over TLS in the XDM-3 interface [4, 13]. Therefore, in general the XDMC will use an 'https' URI (instead of an 'http' URI) to access XML documents stored in any XDMS.

In addition to access to the Aggregation Proxy via an XDMC located at a cellular device, the operator may provide other mechanisms to let end users manage their application policies and groups. In this case, the XDM-3 interface may be implemented by other types of clients (e.g. an application running in a laptop PC). Alternatively, in order to ease usability and user friendliness, the operator may allow secure web-based access to the XDM service through any regular browser: the XDM-3 interface may be implemented on behalf of the user by an operator-hosted proxy / gateway.

Figure 3.3 shows different architectural solutions and services to support the XDM-3 reference point.

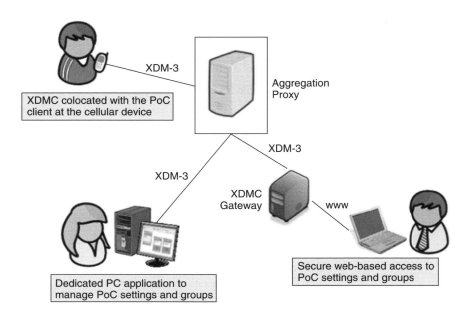

Figure 3.3   Different XDM-3 implementation alternatives

3.3.4 Reference Point XDM-4: Aggregation Proxy – Shared XDMS

The XDM-4 reference point lets the Aggregation Proxy manage XML documents stored in the Shared XDMS on behalf of the XDM client upon user request in a secure way (given that the user has been authenticated). XDM-4 is based on XCAP.

When the XDMC user has been correctly authenticated, the Aggregation Proxy performs a direct mapping of XDM-3 operations into operations against the relevant XDMS (depending on the type of document requested by the XDMC, as explained in section 3.2.3). Whenever the XDMC wishes to perform a request against an XML document stored in the Shared XDMS, the operation is executed over XDM-4 interface. As an example, if an XDMC requests access (XDM-3 interface) to a URI List document, the Aggregation Proxy should redirect this operation towards the Shared XDMS (XDM-4 interface).

Observe that XDM-4 functionality is basically the same as any other reference point connecting the Aggregation Proxy with any XDMS (e.g. PoC XDMS, Presence XDMS, RLS XDMS). The only difference resides in the type of XML document (i.e. the type of XCAP application) involved in each transaction. The Aggregation Proxy uses the XCAP application type (e.g. PoC group, Presence policy, URI List) to route the request to the appropriate XDMS, but the set of operations performed in the interface between the Aggregation Proxy and each XDMS is basically the same.

With this idea in mind, observe that once XDM-4 capability is implemented, it requires only a minor effort for the Aggregation Proxy to support connection to the PoC XDMS (POC-7 interface), Presence XDMS (PRS-12) or RLS XDMS (PRS-12).

3.3.5 Reference Points towards Service-specific Enablers and XDMSs

Communication between the XDM enabler and external entities is supported by reference points defined in enabler-specific documents (e.g. [8, 9, 10, 11]). In general, there are three main functions for which communication beteween the enabler service nodes and the XDMS service nodes is required. These functions are, namely:

* The enabler server must be able to retrieve URI Lists from the Shared XDMS
* The enabler server must be able to retrieve enabler-specific XML documents from the enabler-specific XDMS
* The Aggregation Proxy must be able to create / retrieve / modify / delete XML documents from the enabler XDMS in order to support user's requests.

Table 3.2 provides an overview of the relevant reference points defined by OMA to implement the three functions listed above, in the particular cases of the PoC and Presence enablers.

Even though these functions and reference points are specified in enabler-specific documents [5, 8], observe that the basic capabilities required to implement these reference points are based either on XCAP or on the SIP subscription/notification mechanism. These are the two main mechanisms used in all OMA XDM communications.

In Table 3.2 we see a set of 12 OMA-defined interfaces: POC-5 to POC-8 and PRS-5 to PRS-12. However, we would like to emphasize again that a lot of similar functionality is shared across several interfaces. Observe the following commonalities:

Table 3.2   Reference Points connecting XDM functions and Enabler-specific functions

XDM – Service specific Reference Points					
Endpoints	PoC Reference Point	Presence Reference Point	XDM 'mirror' Reference Point	Most Frequent Operation	Protocol
Enabler Server – Shared XDMS	POC-5	PRS-5, 9		HTTP GET	XCAP
Enabler XDMS – SIP/IP Core[4]	POC-6	PRS-6, 11	XDM-2	–	SIP
Aggregation Proxy – Enabler XDMS	POC-7	PRS-7, 12	XDM-4	–	XCAP
Enabler Server – Enabler XDMS	POC-8	PRS-8, 10		HTTP GET	XCAP

- The logic required to support POC-7 reference point is very similar to the logic required by XDM-4 reference point. In both cases, the XDMS must accept XCAP operations from the Aggregation Proxy, authorize the user and validate that the outcome of the requested operation remains valid within the XCAP framework. Thus, once XDM-4 is implemented, it should be relatively easy to implement POC-7, PRS-7 or PRS-12 reference points.
- POC-5, PRS-5 and PRS-9 reference points do not have a mapping to any XDM reference point. However, this is a relatively simple interface when compared to the previous ones: the enabler server (e.g. the PoC server) never performs any modification in documents stored in the Shared XDMS (these are maintained by the end user). The only operation performed in these interfaces is retrieval of XML documents. Consequently, the complexity of implementing POC-5 (PRS-5, PRS-9) interface should be lower than other interfaces such as XDM-3, XDM-4 or POC-7.
- The above bullet applies in the same manner to POC-8 and PRS-10 reference points. Although [9] indicactes that PRS-8 reference point supports management (create, retrieve, modify, delete) operations over XML documents stored in the Presence XDMS, the only operation specified by [14] is retrieval of XML documents. Therefore, in principle the PRS-8 reference point falls in the same category as POC-8 and PRS-10, unless this contradiction is further clarified by OMA in the future (and relevant use cases are described in which a Presence server requires *write* access to XML documents stored in the Presence XDMS).

3.4 The XML Capability Access Protocol (XCAP)

3.4.1 Introduction

We have seen that the *XML Capability Access Protocol* (XCAP) is a key element on which an integral part of the XDM architecture is based. In order to understand its features and

---

[4] These reference points are to be removed from the corresponding version 1 enabler definitions (PoC, Presence and XDM, respectively), due to the withdrawal of the capability to subscribe to changes in XML documents, as described in section 3.2.2.2.

capabilities, we will provide a brief description of XCAP. The reader is referred to [2, 15] for further details about this topic.

The main purpose of XCAP is to let a client read, write and modify application configuration data (e.g. URI Lists, application settings, service policies) stored in XML format in a XCAP server (e.g. an XDM server). To achieve this goal XCAP defines a set of rules that let a client address XML documents and/or components within XML documents (in general: *XCAP resources*), via well-defined URIs. XCAP is typically implemented over regular HTTP, so that XCAP transactions are directly mapped into standard HTTP primitives (e.g. GET, PUT, DELETE)[5].

The next sections briefly introduce XCAP.

3.4.2 XCAP Application Usages

All XCAP resources stored in an XCAP server must be associated to a certain service or application. This approach lets a single XCAP server store different types of XML documents and hence support different XCAP applications. XCAP is not meant to store generic XML documents without a well-defined format, semantics and purpose: each XCAP document must be associated to a clearly identified XCAP application.

In order to achieve this, XCAP Application Usages are defined: whenever a particular service requires having application information stored in the network and made accessible to the user, the XCAP model can be used. In such a case, a new XCAP Application Usage must be created.

The Application Usage must define how this data is to be stored in the XCAP server, which XML format the data must follow, which constraints (if any) should be applied to XML documents defined under that application and which authorization policies should be applied (e.g. which users are allowed to do what – e.g. read, write – on which documents). That is: the application must define the *grammar* (i.e. the XML schema and additional data constraints) and *semantics* (i.e. the meaning) of XML data used under the context of the application.

When a new Application Usage is created, the associated *Application Unique* ID (AUID) must be defined as well. The AUID is simply a unique text string that lets the XCAP server and client clearly identify under which application a given XML document was created and, therefore, which purpose it serves. For example, the AUID value associated to XML documents containing PoC groups is 'org.openmobilealliance.poc-groups'.

Obviously, every type of application must use a different AUID value. However, a common AUID value (and application usage) may serve different but similar purposes. For example, the AUID value 'resource-list' is associated to URI List XML documents. However, two different documents under the 'resource-list' AUID may serve different purposes: one URI List may store the members of a PoC group, and another URI List may represent the buddies in a Presence subscription list.

XCAP defines a default application that must be supported by any XCAP compliant server, namely: the *XCAP Server Capabilities application*, which uses AUID value 'xcap-caps'. This

---

[5] In fact, XCAP is generally executed over HTTPoverTLS, although this difference is transparent to the application layer (e.g. primitives, operations, headers, . . .), the only difference being the TCP port associated to the service (443) and the 'https' URI [3.2][3.13]

application stores a single XML document, which lets any XCAP client query the server about the list of XCAP applications it supports. The XCAP Server Capabilities application is therefore a 'pointer' to the 'real' XCAP applications supported by the XCAP server. An XCAP server only supporting the 'xcap-caps' AUID is, therefore, a useless server.

Default AUID values defined within the XCAP framework, under the IETF umbrella, have values such as 'xcap-caps' or 'resource-lists'. These general XCAP applications have an AUID value that is defined in the global IETF namespace for XCAP AUID's.

If a given organization wishes to define new XCAP applications, it must do so under the vendor-specific namespace. For such purpose, XCAP mandates that the AUID must be constructed using the reverse domain name of the organization creating the AUID, followed by a period, followed by any organization defined token. OMA-defined XCAP applications fall within this category. Observe that the OMA PoC Groups XCAP application follows this structure ('org.openmobilealliance.poc-groups').

3.4.3 URI Construction in XCAP

XCAP fulfils the key function of addressing XML documents and XML components via well-defined HTTP URIs. In order to manage XML documents properly, XCAP must define an HTTP URI construction mechanism that ensures that any XML component is addressable for any user who has the right to access it. The URI construction mechanism specified by [2] defines how this requirement is achieved.

The XCAP URI performs a logical map between the way information is structured, and the string that defines the URI. In general, an XCAP URI is constructed by concatenating up to four components: the *XCAP root*, the *document selector*, the *node separator* and an optional XML *node selector*. The following example XCAP URI consists of the XCAP root (underlined) and the document selector (no node selector is used in this case):

```
http://xcap.wirelessfuture.com/services/
org.openmobilealliance.poc-groups/users/sip:joe.
doe@wirelessfuture.com/friends.xml
```

The process to construct a valid XCAP URI is described below. All XCAP entities (clients and servers) must be able to construct a valid XCAP URI if they wish to manage XML documents. Thus, the XDMC, the PoC Server, the Presence Server and the RLS must be able to perform the operations described below.

Before explaining the XCAP URI construction process in detail, it is important to understand that this mechanism is 'hierarchical': resources are identified by their logical situation within a logical tree that contains all XCAP resources stored in a given XCAP server. In fact, we will talk about a structure of folders and files within a server's XCAP tree.

However, it is important to understand that this is only a *logical* structure. Given that XCAP is implemented over HTTP, it is perfectly feasible to implement an XCAP server adapting an HTTP server and mapping the XCAP hierarchy to the server's actual folder structure within the Operating System. However, this is an implementation issue: XCAP makes no assumption about how XCAP resources are actually stored in a given implementation, so other alternatives are perfectly feasible (e.g. relational database).

Table 3.3 Percent encoding of reserved characters to produce valid URIs

ASCII character	Percent-encoded sequence
[	%5b
]	%5d
'	%22

In the URIs we use for example purposes in the rest of this chapter, we will not show the right percent encoding defined in [16] for valid URIs. This is to ease URI readability. Table 3.3 provides the mapping of the percent-encoding sequences that should be used in a real XCAP implementation.

3.4.3.1 XCAP Root

The first part of the XCAP URI points to the root of the XCAP tree where XML resources are stored. XCAP recommends that the XCAP root take the form 'xcap.domain'. Therefore, an example valid XCAP root URI would be `http://xcap.wirelessfuture.com`. It is quite frequent, however, that implementations decide to add a `/services` folder after the domain name or FQDN.

In general, the XCAP root URI is provisioned in the XCAP client, as a prerequisite to access XCAP services. This is the case of the XDMC as well. In the particular case of OMA XDM, the XCAP Root URI points to the Aggregation Proxy.

3.4.3.2 Document Selector

The part of the XCAP URI located after the XCAP root is the Document Selector. The Document Selector lets the XCAP client address a given XML document (this is the most general XCAP resource that can be managed via XCAP). The Document Selector is constructed by concatenating a sequence of *path segments* separated by a slash ('/'). These path segments are, namely, the Application Unique ID, the XCAP User Id and the Document Name. They are briefly presented below:

- The Application Unique ID (AUID) is a key element in the XCAP hierarchy, which lets the server determine the application usage of every XCAP document. As an example, an XCAP document containing a list of URIs uses the AUID value `resource-lists`.
- The second part of the document selector indicates whether a given resource is a global one (i.e. applicable to all users of a given application) or is associated to a defined user. There are two default sub-trees defined below the AUID: the `global` tree and the `users` tree.
- When an XML document applies to a given user, the sub-tree below the users tree consists on a username that clearly identifies the user under the XCAP service: the so-called *XCAP User Identity* (XUI). OMA XDM further specifies that the XUI must consist on either the public SIP URI or a TEL URI assigned to the end user that owns the XCAP resource. In

most cases, the XUI is the SIP service identity (i.e. the XUI will generally be equal to the PoC identity).

- Finally, the last part of the document selector is the XML document name, which refers to a document available for the specified AUID and the user identified by the XUI (or applicable to all users, if under the global tree).

If we take the following document selector as an example:

```
/org.openmobilealliance.poc-groups/users/sip:joe.
doe@wirelessfuture.com/friends.xml
```

we observe that the associated XCAP AUID is `org.openmobilealliance.poc-groups` (thus, it contains a PoC group XML document). The document exists within the users tree, and the owner of the document is `sip:joe.doe@wirelessfuture.com`. The document name is `friends.xml`.

It is important to note the difference between global and users documents. For each AUID, documents stored under the `users` folder are owned and managed by end users, generally making use of the XDMC capabilty at their handset. The owner of the document is the user under whose identity the document is stored. In the example above, Joe (`sip:joe.doe@wirelessfuture.com`) is the owner of the `friends.xml` document.

In contrast, documents stored under the global tree are defined at the application level: they are not associated to any particular user. In general, global documents are managed by an application server (e.g. the PoC XDMS, the Shared XDMS), and users cannot modify the contents of a global document.

Observe that XCAP URI creation is a key procedure in the whole XCAP/XDM framework: by implementing a structured process to create the URI, the XDMC is able to manage all of XML documents stored in different XDMSs belonging to a certain user.

XCAP and OMA XDM specify a well-defined combination of user documents managed via XDMC and global documents managed by XMDSs. With the collaboration of several XCAP applications and usage of global and user documents, the XCAP / XDM framework helps users keep track of their XML documents, avoiding the possibility that some information or document gets 'lost' within the XCAP tree. If an XML document exists in the XCAP tree and the user has the rights to manage it, she should be able to retrieve it successfully. Further details about this XCAP feature are provided in section 3.6, where we describe all XCAP applications that are relevant for OMA XDM.

3.4.3.3 Node Selector

Beyond the document selector, XCAP defines how the XCAP URI can be used used to point to specific XML nodes located within a document. XML nodes are elements within an XML document or attributes of an element within an XML document. The XCAP URI lets an XCAP client manage (e.g. create, read, modify or delete) information contained in an XML document without the need to download the full file. In order to ease parsability, the *document selector* and the *node selector* are separated by the *node selector separator*: a double tilde string ('~~').

The node selector is not required for XCAP operations performed at the document level. When the node selector is present in a XCAP URI, the complete URI has the following structure:

```
http://<xcap root uri>/<document selector>~~/<node selector>
```

The detailed procedure to construct the node selector part of the XCAP URI is described in [2]. For our purposes, it is enough to understand that a hierarchical approach is followed, so that elements within an XML document can be seen as 'folders' within the XCAP tree. Let us consider the following XML example document (a PoC group) to present how the node selector works:

```
<?xml version="1.0" encoding="UTF-8"?>
<group xmlns="urn:oma:xml:poc:list-service"
       xmlns:rl="urn:ietf:params:xml:ns:resource-lists"
       xmlns:cr="urn:ietf:params:xml:ns:common-policy"
       xmlns:xsi="http://www.w3.org/2001/XMLSchema-
       instance">
  <list-service uri="sip:my_squash_colleagues@
  wirelessfuture.com">
    <display-name xml:lang="en-us">Squash Colleagues
    </display-name>
    <list>
      <entry uri="tel:+43012345678"/>
      <entry uri="sip:mary.watson@wirelessfuture.com"/>
    </list>
    <cr:ruleset>
      <cr:rule id="a7c">
        <cr:conditions>
          <is-list-member/>
        </cr:conditions>
        <cr:actions>
          <allow-anonymity>true</allow-anonymity>
        </cr:actions>
      </cr:rule>
    </cr:ruleset>
  </list-service>
</group>
```

The XCAP URI that points to the location of this XML document may look like:

```
http://xcap.wirelessfuture.com/services/org.
openmobilealliance.poc-groups/users/sip:joe.doe@
wirelessfuture.com/squash.xml
```

Now, let us assume that Joe wishes to access the <list> element of this XML document. In this case, the node selector part of the XCAP URI should look something like:

```
/group/list-service/list
```

Thus, the complete XCAP URI would be:

```
http://xcap.wirelessfuture.com/services/org.
openmobilealliance.poc-groups/users/sip:joe.doe@
wirelessfuture.com/squash~~/group/list-service/list
```

An HTTP GET operation performed on the above URI would return the following XML content:

```
<list>
  <entry uri="tel:+43012345678"/>
  <entry uri="sip:mary.watson@wirelessfuture.com"/>
</list>
```

Now, let us assume that Joe wishes to find whether he has included Mary in the list of group members. In this case, Joe's XDMC would trigger a search for an <entry> element including a uri attribute equal to Mary's PoC address. This logic can be implemented in XCAP by making use of the following syntax for the codification of the node selector:

```
/groups/list-service/list/entry[@uri="mary.
watson@wirelessfuture.com"]
```

The full URI would look like:

```
http://xcap.wirelessfuture.com/services/org.
openmobilealliance.poc-groups/users/sip:joe.doe@
wirelessfuture.com/squash~~/group/list-service/list/
entry[@uri="mary.watson@wirelessfuture.com"]
```

From the user's point of view, an HTTP GET operation performed against this XCAP URI would mean something like: 'I have a PoC group document including my squash colleagues. I would like to see if I included Mary in that group'. The HTTP GET operation will return the following result:

```
<entry uri="sip:mary.watson@wirelessfuture.com"/>
```

3.4.3.4 Example Usage

Let us assume that Joe has subscribed to a service provider that operates in the domain `wirelessfuture.com`. The XCAP root URI is `http://xcap.wirelessfuture.com/services`.

Joe wants to create a new document in the Shared XDMS to store the list of all his PoC contacts in a document called `index`. Storing the actual PoC contact list in the network lets Joe easily share this list among different PoC devices, and even upgrade his PoC handset without having to manually migrate the contact list from the old to the new phone.

Under these assumptions, the XCAP URI that would let Ron manipulate this document would be:

```
http://xcap.wirelessfuture.com/services/resource-list/users/
sip:ron@wirelessfuture.com/index
```

The XDM client has been able to construct the XCAP URI from a set of well-known pieces:

- `http://xcap.wirelessfuture.com/services` is the XCAP root URI provisioned in the XDMC
- `users/resource-list/sip:joe.doe@wirelessfuture.com/index` is the document selector, where:
  - ○ `resource-list` is the AUID.
  - ○ `users` indicates that the document is stored under the users tree of the AUID.
  - ○ `sip:ron@wirelessfuture.com` is the XUI, which may be provisioned in the XDMC or stored in a SIM/ISIM module attached to the cellular handset.
  - ○ `index` is the standard document name to store shared lists, as defined in [6].

Imagine that, in addition to the creation of the buddy list stored in the Shared XDMS, Joe wishes to define a new PoC group XML document. The process for creating the XCAP URI would be very similar to the one described above, the only differences being the AUID value used for PoC groups (`org.openmobilealliance.poc-groups` instead of `resource-lists`) and the actual document name, which can be decided by Joe.

An example XCAP tree structure of the Shared XDM server operated by wirelessfuture.com is shown in Figure 3.4.

The Shared XDMS supports two types of applications: storage of *URI Lists* and storage of *Group Usage Lists* (i.e. documents that let clients define the type of usage that each PoC group has – e.g. chat, pre-arranged group). These two types of applications use AUID values `resource-lists` and `org.openmobilealliance.group-usage-list` respectively. Under the AUID tree, documents may be stored in the `global` or the `users` tree. We see that in the example below one XML document is stored for each AUID in the `global` tree. Harry and Ron have created some XML documents containing URI Lists (Harry has created three documents and Ron has created only one). Finally, Ron has also stored another XML document under the `org.openmobilealliance.group-usage-list` AUID.

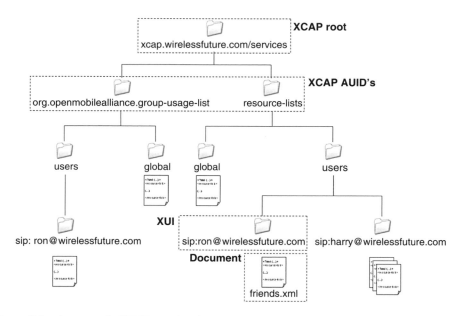

Figure 3.4   An example XCAP tree for the Usage List and the Resource List XCAP applications

Table 3.4   Mapping XCAP operations into HTTP primitives

XCAP Operation	HTTP Primitive
Document / Element / Attribute creation	HTTP PUT
Document / Element / Attribute modification	HTTP PUT
Document / Element / Attribute retrieval	HTTP GET
Document / Element / Attribute deletion	HTTP DELETE

3.4.4 Client-server Communication

An XCAP client is an HTTP/1.1 compliant client as well. Thus, performing XCAP operations consists in mapping them into the right HTTP primitive, using the right XCAP URI and correctly setting some HTTP headers.

Table 3.4 shows how typical XCAP operations map into HTTP primitives. It is worth noting that, apart from the trivial mapping presented there, proper URI generation and XML data validation are key to consistently work with XCAP documents.

When a client is interested in performing an XCAP operation it selects the proper HTTP request. In case of document/node creation or update, the XML client should ensure that the resulting document remains consistent with the data constraints imposed by the application usage. In particular, an XCAP server must not allow any modification that breaks any data constraint, and an XCAP client must not intend any modification that would lead to issues with the application-defined schema and constraints. The following items must be checked by an XDM Client and Server before proceeding with a change:

- After modification, the document must remain well formed and fully compliant to the XML schema defined by the application usage.
- Clients and servers must make use of the proper MIME-Type when exchanging XML components over XCAP. Each XCAP AUID defines the associated MIME-Type, thus XCAP requires clients and servers to be specific and not use the generic `application/ xml` MIME-Type used for plain XML documents.
- All data constraints defined by an application apply after modifying a document. For example, a URI List document may specify that each URI is unique in the context of a given list (i.e. a SIP address cannot be repeated in a URI List). Additionally, applications may define URI constraints (e.g. elements in a URI List may be of type SIP URI or TEL URI, but not any other URI type).
- In certain cases, a client may make use of HTTP capabilities for conditional operations. By making use of the HTTP `If-Match` or `If-None-Match` headers, a client may request that certain operations (e.g. insertion of a new element) be performed only in case that a precondition is met.

Finally, it is worth noting that for all client-server successful transactions (i.e. those concluded with a 2xx response) the server must include an `Etag` header. This header contains a random string which is used for identification purposes. Whenever any content within an XML document changes a new `Etag` value must be computed and delivered by the server. This mechanism lets the client perform conditional requests or validate whether locally stored copies of XML documents are up-to-date or not.

3.5 User Authentication and Authorization

3.5.1 Introduction and User Identity

User authentication consists on determining whether a user wishing to perform certain operation is who he or she claims to be. User authorization consists on determining whether an authenticated user is allowed to perform the intended operation.

In the OMA architecture, the Aggregation Proxy is responsible for User Authentication. User authorization is performed by the relevant XDMS that processes requests coming from authenticated users.

User access to XCAP services must be performed in a secure way. For SIP-based services, authentication of the signalling flow is performed by the SIP/IP Core. However, XCAP is an HTTP-based service, which does not cross the SIP/IP Core infrastructure. The Aggregation Proxy fills this gap by taking responsibility over user authentication tasks.

The fact that XCAP supports user authentication indicates that the concept of XCAP User Identity (XUI) exists. Observe that any access to XCAP services must be performed by an authenticated user. In the following sections, we describe XDM authentication and authorization functions, as well as XCAP traffic encryption as defined by [2, 4]. When XCAP services are used in the PoC framework, the XUI is generally equal to the PoC identity of the end user that owns the PoC client and the XDMC. Therefore, the XUI generally takes to form of a TEL URI or a SIP URI.

3.5.2 User Authentication and XCAP Traffic Security

The Aggregation Proxy is the entity responsible for user authentication. In fact, the Aggregation Proxy effectively behaves as an HTTP Proxy (i.e. the XCAP root URI points to the

Aggregation Proxy). The default authentication mechanism specified by XCAP and XDMS is HTTP Digest [17]. Thus, the Aggregation Proxy must use HTTP Digest to challenge the client and authenticate it.

The HTTP Digest mechanism can only be used with XCAP when Transport Layer Security is activated [2]. By combining HTTP Digest mechanism and TLS encryption, XCAP communications between the XDMC and the Aggregation Proxy are protected against eavesdropping

Usage of HTTP Digest poses a requirement to provision XCAP clients with username and password parameters. These parameters are used when challenged by the Aggregation Proxy. In order to avoid this restriction and to leverage well-known authentication information available to 3G devices (e.g. stored in the USIM/ISIM applications), OMA may reuse the 3GPP *Generic Authentication Architecture* (GAA) [18] [19] when deployed over a 3GPP network supporting this capability. This functionality is explained in further detail in chapter 5.

In order to let the XCAP/XDM server know the authenticated identity of the requesting user, the Aggregation Proxy inserts an HTTP header in all requests being proxied from a successfully authenticated client. When regular HTTP Digest is used to authenticate the user, a X-XCAP-Asserted-Identity is inserted, while the X-3GPP-Asserted-Identity header is inserted when authentication takes place using 3GPP GAA. The following examples show how these headers could look:

```
X-XCAP-Asserted-Identity: "sip:joe.doe@wirelessfuture.com"
X-3GPP-Asserted-Identity: "sip:joe.doe@wirelessfuture.com"
```

Observe that both headers will never be present in the same HTTP message, given that they are inserted when two mutually exclusive authentication mechanisms are used.

As a result, having authentication responsibility assigned to the Aggregation Proxy lets the Application Server (e.g. the XCAP/XDM server) assume that all incoming HTTP requests come from either authenticated users or network servers located in a trusted domain (e.g. the PoC Application Server). Another important consequence is that all communications between the Aggregation Proxy and each XDMS do not need to use TLS, and they can run regular HTTP over TCP, thus relaxing the processing power required to run the XDMS logic (encryption is terminated by the Aggregation Proxy).

Whenever user authentication fails, the request is directly rejected by the Aggregation Proxy with a 403 Forbidden answer.

Observe that 3GPP discourages usage of HTTP Digest as an authentication mechanism for HTTP-based services such as XCAP, so the reader interested in deployment of XCAP services in 3GPP environments is referred to chapter 5, where additional details about GAA and *Early Ut* Authentication,[6] as defined by 3GPP, are provided.

Finally, it is worth observing that OMA XDM authentication works slightly differently from IETF default XCAP authentication. [2] specifies that XCAP servers must be able to challenge XCAP clients using HTTP Digest means. This is not required anymore for XDM servers, since it is the Aggregation Proxy challenging the XDMC (thus, XDMSs need not

---

[6]Early Ut authentication is a mechanism defined by 3GPP to enable authentication of HTTP-based IMS services for devices that are not equipped with an ISIM application. Ut is a 3GPP-defined interface, that effectively maps into OMA XDM-3 reference point. Refer to chapter 5 for further details.

support HTTP Digest or TLS). From this point of view, someone could argue that an XDM server is not a fully compliant XCAP server. Rather, it is the logical combination of the Aggregation Proxy and the XDMS functions that build up an XCAP compliant server.

3.5.3 User Authorization

XCAP defines a default authorization rule for XCAP documents: a user has full access (i.e. create, read, write, modify) to all documents stored under her home directory (i.e. the one named after the user's XUI). Additionally, all users can read all XCAP documents located anywhere under the global directory. Every XCAP application may – if needed – modify the default authorization policy defined by XCAP [2].

XDMS defines a slightly different behaviour for the default authorization policy: XDM servers will grant access (read/write) permission only to owners of XML documents. Therefore, if user joe.doe@wirelessfuture.com has a set of XML documents stored through different XDM servers (e.g. Shared XDMS, PoC XDMS, Presence XDMS and/or RLS XDMS) he will be the only user allowed to retrieve those documents and make modifications on them.

Regarding documents stored under the global tree, [4] mandates that each application must explicitly define the authorization rule to be applied by the XDMS. Thus, no default rule is defined by [4]. This is in contrast with plain XCAP [2] default authorization rule, described above.

Given the simplicity of the User Authorization policies specified by OMA XDM, a simple match between the document owner identity (contained in the XCAP URI of a request) and the authenticated identity of the requesting user (contained in either the X-XCAP-Asserted-Identity or in the X-3GPP-Asserted-Identity header) lets the XDM Server determine whether a user is authorised to perform the requested XCAP operation or not.

3.5.4 Authentication and Authorization Sample Signalling Flow

As a summary of all XCAP/XDM authentication and authorization mechanisms, we will use the sample signalling flow presented in Figure 3.5 to present how the overall procedure works.

In this case, Joe wishes to retrieve a URI list stored in the Shared XDMS. Let us assume that the XCAP URI of the HTTP GET request is:

```
http://xcap.wirelessfuture.com/services/resource-list/users/
sip:joe.doe@wirelessfuture.com/index.xml
```

The process to retrieve this document, including authentication and authorization mechanisms is as follows:

1. The XDMC sends an initial HTTP GET request to the Aggregation Proxy. It includes a X-3GPP-Intended-Identity header to indicate the identity of the user that has initiated the request.

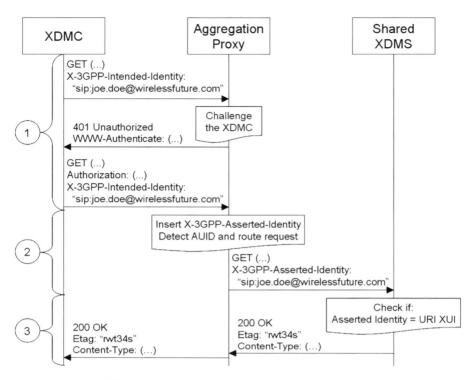

Figure 3.5 Sample Authentication and Authorization signalling flow

Since the request comes from a yet unauthenticated user, the Aggregation Proxy challenges this request by sending a 401 Unauthorized response that includes a WWW-Authenticate header with the HTTP Digest challenge.

The client finally sends an HTTP GET request with an Authorization header that contains the response to the challenge, as per HTTP Digest [17].

2. At this stage, the Aggregation Proxy determines that the user has been correctly authenticated, it checks the AUID value of the XCAP URI (resource-lists) and decides that the request be proxied towards the Shared XDMS. The Aggregation Proxy inserts a X-3GPP-Asserted-Identity to indicate that the user has been successfully authenticated.

3. The Shared XDMS detects that the user tries to access a document in the users tree. The authorization policy mandates that the only user authorized to retrieve such document is the owner of the document. The owner of the document is included as the XUI part of the XCAP URI (sip:joe.doe@wirelessfuture.com).

The Shared XDMS looks for the X-3GPP-Intended-Identity and detects that the requesting user is the same as the owner of the document, and thus authorizes the request. It sends back a 200 OK response containing the XML document requested by Joe. This response has the following characteristics:

- It includes a Content-Type: application/resource-lists header to indicate the specific content type

- It includes an `Etag: 'rwt34s'` header, to let the client keep track of the document version
- It contains the `index.xml` document.

Finally, the response is proxied back to the client by the Aggregation Proxy.

Observe that communication in the XDMC – Aggregation Proxy path uses TLS transport. The transactions explained above occur for the initial transaction while the XCAP user has not been authenticated. If the same XDMC intends to perform a subsequent XCAP operation, it will not be challenged by the Aggregation Proxy (unless an inactivity timer expires for this connection before the second operation is executed).

3.6 XCAP Applications and Documents Used in OMA XDM

3.6.1 Introduction

We have seen that XCAP is a generic approach to management of application configuration documents. From this point of view, the way XCAP/XDM works is exactly the same independently of the type of application or server (Shared XDMS, PoC XDMS, Presence XDMS, RLS XDMS). Although servers are involved in data validation and constraint enforcement – as far as defined by a given XCAP application – the XCAP/XDM servers are purposely unaware of application data semantics, which are for the final application (e.g. PoC, Presence) to interpret. This approach enables high flexibility and extensibility to the XDM architecture to support services that are even yet to be conceived.

Therefore, we expect definition of new XCAP applications in the future, as new services are developed (e.g. PoC v2.0, SIMPLE Messaging, Presence v2.0). The number and applicability of XCAP services will grow over the time. In this section, we present all XCAP applications defined at the time of writing (September 2007) by version 1 of the PoC, Presence and XDM OMA enablers.

3.6.2 Common XCAP Applications Supported by all XDM Servers

3.6.2.1 XCAP Caps

The first XCAP application usage supported by any XCAP compliant server uses the AUID value `xcap-caps`. This application defines a single XML document named 'index' be stored in the global tree, which describes all XCAP capabilities supported by the server.

The global document defined by the `xcap-caps` application stores three sets of information about the server, namely:

- the list of XCAP applications (AUID's) supported by the server
- the list of supported extensions – if any – of each AUID
- and the list of supported XML namespaces (at least, those which correspond to the supported XCAP applications must be listed).

Given the nature of the `xcap-caps` service, all XCAP users must have 'read' rights to the global document listing XCAP capabilities.

The `xcap-caps` global document lets the client uniquely construct the XCAP URI required to retrieve the XCAP capabilities of the server. Let us assume that the XCAP root URI is provisioned in a XCAP client is `http://xcap.wirelessfuture.com/ services`. In this case, the XCAP URI that will let the client retrieve the xcap-caps main document is:

```
http://xcap.wirelessfuture.com/services/xcap-caps/index
```

When a client sends an `HTTP GET` request towards this XCAP URI, the server returns an XML document like the one shown below.

```
<?xml version="1.0" encoding="UTF-8"?>
<xcap-caps xmlns="urn:ietf:params:xml:ns:xcap-caps"
        xmlns:xsi="htt//www.w3.org/2001/XMLSchema-instance"
        xsi:schemaLocation="urn:ietf:params:xml:ns:
        xcap-caps xcap-caps.xsd">
    <auids>
        <auid>xcap-caps</auid>
        <auid>resource-lists</auid>
        <auid>org.openmobilealliance.xcap-directory</auid>
        <auid>org.openmobilealliance.group-usage-list
        </auid>
    </auids>
  <extensions>
        <!- No extensions supported ->
  </extensions>
  <namespaces>
        <namespace>urn:ietf:params:xml:ns:
        xcap-caps</namespace>
        <namespace>urn:ietf:params:xml:ns:
        xcap-error</namespace>
        <namespace>urn:ietf:params:xml:ns:resource-lists
        </namespace>
        <namespace>urn:oma:params:ns:resource-list:
        oma-uriusage</namespace>
  </namespaces>
</xcap-caps>
```

This document indicates that the server supports four XCAP applications: `xcap-caps`, `resource-lists`, `org.openmobilealliance.xcap-directory` and `org. openmobilealliance.group-usage-list`. All the namespaces required to support these XCAP applications are listed in the `<namespaces>` element.

The `xcap-caps` application lets the client know what type of XCAP server it is communicating with and determines if it will support the XCAP operations it intends to perform. Any client may query an XCAP server to determine the server's capabilities. Observe that this information is required by the client to construct valid XCAP URIs and to perform XCAP operations under the rest of applications supported by the server.

In the OMA XDM architecture, whenever the client queries the `xcap-caps` application, the Aggregation Proxy will fork the HTTP GET request to all XDMSs, and it will deliver to the client a response aggregating all `xcap-caps` documents from the different servers. The XDM architecture looks to the client as a single XCAP server (regardless of the actual implementation of the different servers and interfaces), due to the consolidation capabilities of the Aggregation Proxy.

Observe that the `xcap-caps` document shown above may represent the capabilities of the Shared XDMS [6]. Therefore, this document 'as is' will generally not reach the XDMC: it would be delivered by the Shared XDMS to the Aggregation Proxy. The Aggregation Proxy will merge this document with the rest of `xcap-caps` documents coming from other XDMSs, and deliver a single consolidated `xcap-caps` document to the XDMC, conveying all capabilities from the Shared XDMS, the PoC XDMS, the Presence XDMS and the RLS XDMS.

3.6.2.2 XML Documents Directory

This is an OMA XDM specific application (a generic XCAP server does not implement this application). The XML Documents Directory feature lets a user keep track of all XML documents that are stored in all XDM servers. It also lets the XDMC retrieve a list of all XML documents stored under a particular application. For example, the user could be interested in retrieving all PoC Group documents that she has created in the PoC XDMS: the XML Documents Directory application lets the user obtain that list.

This application uses AUID value `org.openmobilealliance.xcap-directory`, and it must be supported by all XDM compliant servers. Each XDMS is responsible for creating and maintaining a `directory.xml` document for each XUI that it serves. Users are only allowed to retrieve information contained in that document, but not to modify it. The server automatically updates this document, if needed, when the user creates or removes an XML document.

There is a well-known way to retrieve the `directory.xml` document. For example, Joe (`sip:joe.doe@wirelessfuture.com`) may access his directory of XML documents stored in all XDM Servers by using the following XCAP URI:

```
http://xcap.wirelessfuture.com/services/org.
openmobilealliance.xcap-directory/sip:joe.
doe@wirelessfuture.com/directory.xml
```

When a user wishes to retrieve a list of all XML documents stored in all XDMSs, the XDMC sends a request towards the Aggregation Proxy using the XCAP URI presented above as the Request URI. The Aggregation Proxy will query each XDMS individually and consolidate an aggregated answer towards the client.

This application helps the client consistently access XML documents without the need of caching and remembering the hierarchy and names of all XML documents stored at each XDMS.

In addition to the general query to request all XML documents, the client may use this application to retrieve all XML documents stored under a particular XCAP application (e.g. all URI Lists stored at the Shared XDMS). In order to achieve this, the client extends the URI with the proper node selector indicating the application being requested. The Aggregation Proxy will then only query a single XDMS to fulfil the request.

The example below shows a sample XML Documents Directory document. It lists the documents owned by Joe, who happens to have one URI list document stored in the Shared XDMS (index) and two PoC group documents stored in the PoC XDMS (skiing and shopping respectively).

```
<?xml version="1.0" encoding="UTF-8"?>
<xcap-directory   xmlns="urn:oma:params:xml:ns:
xcap-directory"
                  xmlns:xsi="http://www.w3.org/2001/
                  XMLSchema-instance">
  <folder auid=resource-lists>
    <entry uri="http://xcap.wirelessfuture.com/services/
      resource-lists/users/sip:joe.doe@wirelessfuture.com/
      index"
        etag="pqr999"/>
  </folder>
  <folder auid=poc-groups>
    <entry uri="http://xcap.wirelessfuture.com/services/
      org.openmobilealliance.poc-groups/users/
      sip:joe.doe@wirelessfuture.com/skiing" etag="abc123"/>
    <entry uri="http://xcap.wirelessfuture.com/services/
      org.openmobilealliance.poc-groups/users/
      sip:joe.doe@wirelessfuture.com/shopping"
      etag="def456"/>
  </folder>
</xcap-directory>
```

If Joe wants to retrieve a list of all PoC groups stored in the PoC XDMS, he only needs to send an HTTP GET message to the following URI:

```
http://xcap.wirelessfuture.com/services/org.
openmobilealliance.xcap-directory/sip:joe.doe@
wirelessfuture.com/directory~~/xcap-directory/folder
[@auid="poc-groups"]
```

Such a request would return the result of searching for a folder element within the directory document, containing an auid attribute with value poc-groups (i.e. the <folder auid=poc-groups> element).

3.6.3 XCAP Applications Supported by the Shared XDMS

3.6.3.1 URI Lists

The main goal of the Shared XDMS is to store lists of URIs. These lists may be reused by enabler- specific XDMSs to construct application-specific groups whose members are the entries in the list. This way, a common list of contacts may be reused through several applications

Apart from common mandatory XCAP and XDM applications (XCAP Capabilities and XML Documents Directory), the URI List capability is the only mandatory application that must be supported by the Shared XDMS.

URI Lists are fully compliant to the IETF defined Resource List XCAP application [7]. Hence, OMA URI List documents use the `resource-lists` AUID value defined by IETF [7].

A single XML document may contain several URI Lists (each list is defined in a different `<list>` element with a different name attribute). [6] mandates that a single XML document named `index` be used to store all URI Lists of a particular user. This convention eases XCAP operations, since the document containing URI lists can always be retrieved using a well-known URI. Since each list is assigned a different name attribute, the XDMC has the ability to distinguish among different lists.

URI Lists may become complex if nested lists (lists of lists) are used. Therefore, OMA recommends that URI Lists should avoid references to external lists. OMA defines additional extensions to the URI lists schema defined in [7] to extend basic IETF resource list functionality. The reader is referred to [6] [7] for further information.

OMA has defined four default list names, which can be used by XDMC's to define four particular types of lists. These URI names are:

- The `oma_allcontacts` list. This name can be used by the XDMC to store the list of all contacts a user has. These contacts may or may not be associated to OMA services. For example, a cellular device could decide to store the general contact list under this name.
- The `oma_buddylist`, if present, shall be used to store a list of URIs that is associated to any OMA service.
- The `oma_pocbuddylist` can be used to store the PoC contact list.
- The `oma_blockedcontacts` is used to define a list of blocked users. Applications may refer to this list to create service-specific blacklists. As an example, the PoC application may refer to the 'oma_blockedcontacts' list to enable a reject list, and the Presence application may refer to the 'oma_blockedcontacts' to define the list of users who are not allowed to receive Presence information about the owner of the group. What 'blocked' actually means is defined by each application, but the list of blocked users can be shared among several services.

These default lists let the user accesses OMA services from different devices (e.g. a cellular phone, a laptop, a PDA) over time. When each client device is started it may retrieve some well-known documents and lists (such as the 'oma_pocbuddylist' list). This is an implicit synchronization mechanism, which keeps all devices up-to-date without the user having to care about porting information across devices.

The example below shows a sample URI list XML document.

```
<?xml version="1.0" encoding="UTF-8"?>
<resource-lists xmlns="urn:ietf:para:xml:ns:resource-lists"
                xmlns:oau="urn:oma:xml:xdm:resource-list:
                appusage">

 <list name="oma_pocbuddylist">
  <list name="close-friends">
    <display-name>Close Friends</display-name>
    <entry uri="sip:mary.watson@wirelessfuture.com">
     <display-name>Mary</display-name>
    </entry>
    <entry uri="tel:5678;phone-context=+43012349999"/>
  </list>
    <entry uri="sip:joe.blogs@wirelessfuture.com">
     <display-name>Joe</display-name>
    </entry>
 </list>

 <list name="oma_blockedcontacts">
 </list>
</resource-lists>
```

The above URI list document contains two <list> elements. The first one is the OMA PoC buddy list, and it is built up of one <list> sub-element, and one additional URI etry. Furthermore, the document contains an empty list of OMA blocked contacts (the owner of this list has not enabled any blocked contact yet).

3.6.3.2 Group Usage Lists

The Shared XDMS may optionally support the Group Usage List application. This application uses the AUID value org.openmobilealliance.group-usage-list. The client may query the xcap-caps application to determine whether the Shared XDMS supports this optional feature or not.

The Group Usage List application lets the XDMC and the PoC server know, for each group that a user has created, what type of group it is (Prearranged PoC group or a Chat PoC group).

The Shared XDMS may store a single document (named index) for each XUI under the Group Usage List application. That document lists all PoC groups created by the user and indicates the group type.

Knowing whether a PoC group is of type Prearranged or Chat is important to execute the correct PoC session setup mechanism: in case of a Prearranged PoC group session, the group members must be proactively invited by the PoC server; on the other hand, when a user joins a Chat PoC session the server does not need to invite other participants.

The following XCAP URI may be used by Joe (`sip:joe.doe@wirelessfuture.com`) to retrieve the Group Usage Lists XML document:

```
http://xcap.wirelessfuture.com/services/org.
openmobilealliance.group-usage-list/users/joe.
doe@wirelessfuture.com/index
```

An example Group Usage List document is shown below.

```
<?xml version="1.0" encoding="UTF-8"?>
<resource-lists xmlns="urn:ietf:params:xml:ns:
resource-lists"
    xmlns:xsi="http://www.w3.org/2001/XMLSchema-instance"
    xmlns:ou="urn:oma:params:xml:ns:resource-list:
    oma-uriusage"
    xmlns:opu="urn:oma:params:xml:ns:oma-pocusage">
  <list name="bookmarkedPoCGroups">
    <entry uri="sip:managers_team @wirelessfuture.com">
        <display-name>Joe's golf friends</display-name>
        <ou:uriusages>
              <opu:pocusage>prearranged</opu:pocusage>
        </ou:uriusages>
    </entry>
    <entry uri="sip:technical_team@wirelessfuture.com">
        <display-name>Bob's ski friends</display-name>
        <ou:uriusages>
            <opu:pocusage>chat</opu:pocusage>
        </ou:uriusages>
      </entry>
  </list>
</resource-lists>
```

3.6.4 XCAP Applications Supported by the PoC XDMS

3.6.4.1 PoC Group

The PoC Group application is the most important feature of the PoC XDMS. It uses AUID value `org.openmobilealliance.poc-groups`. PoC groups are XML documents that contain all information required to setup a PoC group communication with other PoC users (i.e. the members of the PoC group). A sample PoC group document is presented below.

```
<?xml version="1.0" encoding="UTF-8"?>
<group xmlns="urn:oma:params:xml:ns:list-service"
xmlns:rl="urn:ietf:params:xml:ns:resource-lists"
xmlns:cr="urn:ietf:params:xml:ns:common-policy">
xmlns:ocr="urn:oma:params:xml:ns:common-policy"
xmlns:xsi="http://www.w3.org/2001/XMLSchema-instance"

   <list-service uri="sip:technical_team@wirelessfuture.com">
      <display-name>Technical Team</display name>
      <list>
         <entry uri="tel:+43012345678"/>
         <entry uri="sip:bruce.weinberg@wirelessfuture.com"/>
         <entry uri="sip:patti.bittan@wirelessfuture.com"/>
      </list>

      <invite-members>true</invite-members>

      <max-participant-count>10</max-participants-count>

      <cr:ruleset>
        <cr:rule id=»a7c»>
           <cr:conditions>
              <is-list-member/>
           </cr:conditions>
           <cr:actions>
              <join-handling>true</join-handling>
              <allow-anonymity>false</allow-anonymity>
              <allow-conference-state>false
              </allow-conference-state>
              <allow-initiate-conferece>true<allow-initiate-
              conferece>
              <allow-invite-users-dynamically>false
              </allow-invite-users-dynamically>
           </cr:actions>
        </cr:rule>
      </cr:ruleset>

   </list-service>
</group>
```

The most important elements that can be found in a PoC Group XML document are

- The document contains the declaration of namespaces in the group root element.
- The document contains a <list-service> element. This element must include a uri attribute, containing the PoC group identity. The PoC group identity must be a valid,

unique, addressable TEL URI or SIP URI. In our example, the group identity is `sip: technical_team@wirelessfuture.com`. The `<list-service>` element contains all required information about the PoC group.

- A `<display-name>` element may be included, to associate a human readable string to the PoC group.
- A `<list>` element must be included. It may contain individual `<entry>` elements identifying each individual member of the group (using the `uri` attribute of the `<entry>` element), or an `<external>` element pointing to a URI List stored in the Shared XDMS. In our example, the PoC group contains a list of three members, without referencing any document stored in the Shared XDMS. A `<list>` element may contain a sequence of `<entry>` and `<external>` elements, if needed.

  In particular, if the group owner knows in advance that the group members list may be relevant for several applications (Presence, Messaging) it is wise to store a URI List in the Shared XDMS and use an `<external>` element to define the list of group members.

- An `<invite-members>` element may be included. This element defines the type of PoC group stored in the document. To define a Pre-arranged PoC group, this element is set to `true`. A Chat group is defined setting this element to `false`. This element has an impact on how group sessions are managed by the server (in case of a Pre-arranged PoC group, the server must invite all group members, while for Chat PoC groups the members proactively join ongoing chat sessions). In the absence of this element, a `false` value (Chat) is assumed.

- A `<max-participant-count>` element may be included. When the number of participants in a group session reaches this value, no further users can join the session. Observe that the list of members of the group may be higher than this value, but in such case only the first `<max-participant-count>` users joining a session will be allowed to participate in it (further participants would be rejected with a '`486 Busy Here`' code.

- A set of rules can be configured by the owner of the group, to govern how the server should manage group sessions. The following settings can be configured:
  - Allow or block anonymous participants (group members hiding their identity).
  - A participant in a PoC session may use the subscription to the *conference event package* [20] feature to know who is participating in a session. The `<allow-conference-state>` element lets the group owner regulate who is allowed to use this capability.
  - The group owner determines who is allowed to initiate PoC group sessions (`<allow-initiate-conference>`).
  - He may decide whether group members are allowed to invite external PoC users to participate in PoC group sessions. When this feature is not enabled, only group members explicitly listed in the XML document can participate in group sessions. Enabling this feature is useful when the owner thinks that, sporadically, inviting external users may be useful.

These group settings let the group owner precisely define the associated PoC group policies. The PoC Server must take into account these policies when a new group session starts. In fact, these group policies are a particular case of the more general policies described in [21] and [4], which are defined by users to manage SIP-based services. In section 3.6.4.1.1, we briefly comment how such policies can be defined.

An important feature of the PoC Group application is the definition of a global document: the PoC XDMS must store a document under the global tree that contains a list of all groups created in the PoC XDMS. This document lets the PoC server easily obtain the <list-service> element that defines any PoC group stored in the PoC XDMS. The XCAP URI of this global document is:

```
http://xcap.wirelessfuture.org/org.openmobilealliance.
poc-groups/global/index
```

This list of all PoC groups available in the PoC XDMS helps the PoC server ensure PoC group name uniqueness when a user intends to create a new PoC group.

PoC Group Sample Operation: Inviting the Members of a Pre-Arranged PoC Group
In this section we describe a sample signalling flow that illustrates how a PoC group XML document can be used when initiating a PoC session. Let us consider that User A (sip:clientA@wirelessfuture.com) has created a Pre-Arranged PoC group document and has stored it in the PoC XDMS via XDM-3 interface. The group is called sip:my_friends@wirelessfuture.com. User A has added three members to the group, namely: User B, User C and User D.

Figure 3.6 shows the actions that are triggered when User A initiates a Pre-Arranged PoC group session with this group.

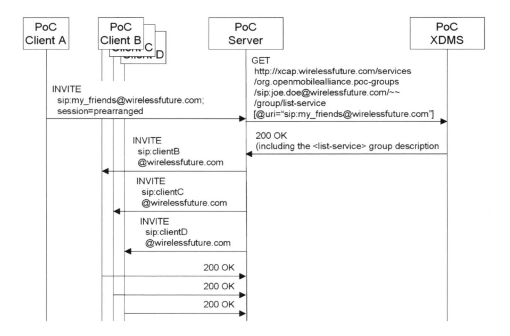

Figure 3.6   Sample signalling flow for PoC Group XCAP operations

1. Initially, User A sends an INVITE request using the group identity as the Request URI of the request. The message is proxied until the PoC server (the SIP/IP Core function is not shown for sake of simplicity).
2. The PoC server detects that this is a Pre-arranged PoC group session (observe the SIP URI parameter), and triggers a request to the PoC XDMS to retrieve the PoC group description, using the mechanism described in section 3.6.4.1. Observe that the structure of the XCAP URI required to retrieve the group information is:

```
http://xcap.wirelessfuture.org/services/
org.openmobilealliance.poc-groups/sip:joe.
doe@wirelessfuture.com/~~group/list-service[@uri="sip:
my_friends.wirelessfuture.com"]
```

3. The result of the GET request delivers the `<list-service>` element that contains all information about the PoC group.
4. The PoC server retrieves the actual list of members of the group and generates one INVITE transaction per group member. For each message, the PoC server maps each entry URI in the list to the Request URI of the INVITE request.
5. Each client may accept or reject the session invitation, and the prearranged PoC group sessions proceeds using standard SIP and PoC mechanisms.

3.6.4.2 PoC User Access Policies: Defining Rules, Policies and Privacy Settings

Introduction
The other application supported by the PoC XDMS is storage of PoC User Access Policy XML documents. The associated AUID value is `org.openmobilealliance.poc-rules`. The purpose of a PoC User Access Policy document is to let users define policies that provide the PoC server guidance on how to serve incoming calls from other users. As an example, a User Access Policy document may be used to set up a blacklist of contacts that should be directly rejected by the PoC server when trying to setup a call against the owner of the blacklist. The PoC User Access Policy XCAP application defines a single document stored under the `users` tree. The PoC Rules document is called `pocrules`. Therefore, it is straightforward for the XDMC to access the PoC Rules document associated to a user by building an XCAP URI with the following structure:

```
http://<XCAP root URI>/users/org.openmobilealliance.
poc-rules/<XCAP User Identity>/pocrules
```

Definition of rules and policies is one important XCAP feature: the goal of any service (PoC, messaging, conferencing) is to offer added value to the end user, avoiding bad user experience such as spam, intrusive communications or malicious usage. In order to achieve this goal, a mechanism exists within the XCAP framework to let users manage service policies and privacy settings [21].

The *Common Policy* mechanism defined in [21] is the basic building block of the PoC User Access Policy application, but it is also a key element of other policy and rule-based applications, such as the *Presence Authorization Rules* and the *Presence Subscription Rules* discussed in chapter 4.

Because of this widespread usage of the Common Policy XCAP framework [21] within the XDM architecture, we dedicate the following section to present PoC Rules documents and policies in some detail.

Policy Documents

In order to let users have enough flexibility when managing PoC policies, OMA relies on an RFC that defines a general framework to define privacy preferences [21]. By combining [4] and [21], a user has a complete set of tools that let her configure privacy policies associated to the PoC service.

A PoC User Access Policy document is compliant to the XML schema defined in [21], with some additional extensions and restrictions specified by OMA [4]. A policy is defined in [21] as a set of 'rules'. Thus, the basic structure of an XML policy document is as follows:

- The root is a `<ruleset>` element which contains all namespace declarations
- The `<ruleset>` element may contain a set of zero or more `<rule>` elements.
- Each `<rule>` element can be thought as an 'if . . . then . . .' statement: it defines what to do in case certain conditions are met (i.e. when they are evaluated to a TRUE value). Therefore, each rule may contain a `<condition>` element (the 'if' part), an `<actions>` and a `<transformations>` element (these build up the 'then' part of the statement).

Several child elements can be used in the `<condition>` part of the rule. The most relevant for us is the `<identity>` element, which is used to match a rule against one or more identities (e.g. PoC addresses). In addition, [4] defines the `<anonymous-request>` and `<other-identity>` elements. By combining these three elements, a user can define privacy rules such as the following ones (the underlined part of the statement highlights the identity-based condition of the rule):

1. *I do not want to let <u>any user from domain 'poc-spammers.com'</u> initiate a PoC session with me*
2. *I want to block <u>users hiding their identity</u> from contacting me.*
3. *I want that whenever <u>Bob (sip:bob.jackson@wirelessfuture.com)</u> tries to PoC me, the server asks me to accept the call using manual answer mode, regardless of my answer mode setting activated at that time.*

As an example, the following condition element includes an `<identity>` child element that implements the logic 'everyone except users from domain *poc-spammers.com*'. When inserted into a complete rule definition (including an `<actions>` element), this rule will be applied to all incoming requests except those coming from this domain.

```
<conditions>
    <identity>
        <many>
            <except domain="poc-spammers.com"/>
        </many>
    </identity>
</conditions>
```

The `<actions>` and `<transformations>` part of the rule (globally known as the 'permissions') are defined outside of [21]. Each application must define what actions and transformations are relevant in each case. For example: for the PoC service, the relevant actions may consist on blocking or allowing incoming sessions, while in the case of Presence, relevant actions may consist on accepting or denying subscriptions to Presence information. This split between the general framework [21] and the application-specific definitions means that policies can never be described completely using solely [21], but they have always to be completed with additional definitions coming from other IETF or OMA specifications.

In particular, PoC actions are defined in [8]. Presence actions and transformations are defined in [22] and [10]. We will describe Presence policy documents in further detail in chapter 4.

An example XML PoC User Access Policies document is shown below.

```
<?xml version="1.0" encoding="UTF-8"?>
<ruleset xmlns="urn:ietf:params:xml:ns:common-policy"
         xmlns:poc="urn:oma:params:xml:ns:poc-rules"
         xmlns:ocp="urn:oma:params:xml:ns:common-policy"
         xmlns:xsi="http://www.w3.org/2001/
         XMLSchema-instance">

  <rule id="f3g44r1">
    <conditions>
      <identity>
        <one id="tel:5678;phone-context=+35812349999"/>
        <one id="sip:mary.anderson@wirelessfuture.com"/>
      </identity>
    </conditions>
    <actions>
      <poc:allow-invite>accept</poc:allow-invite>
    </actions>
  </rule>

  <rule id="ythk764">
    <conditions>
      <ocp:anonymous-request/>
    </conditions>
    <actions>
      <poc:allow-invite>reject</poc:allow-invite>
    </actions>
  </rule>

</ruleset>
```

This document contains two rules: the first one will apply to two particular PoC users, who are included in the acceptance list. Most likely, the owner of the document has good knowledge about these two people and trusts them.

On the other hand, the second rule defines that all requests coming from users hiding their identity be rejected immediately by the server, without notifying the end user. Observe that both rules make use of the `<allow-invite>` element, defined in [4] as a child element of the `<actions>` element. In addition, rule 'ythk764' uses the `<anonymous-request>` element to build up the condition part of the rule.

The three values defined by OMA for the `<allow-invite>` element are:

- `'pass'`. The PoC server will pass the request to the PoC user using the manual answer mode procedure (regardless of the default answer mode configured by the recipient at the time of the request).
- `'reject'`. The incoming call will be rejected by the PoC server without alerting the recipient
- `'accept'`. The PoC server will process the incoming call request using the default answer mode configured by the recipient when the request arrives.

Example Signalling Flow
In this section we present PoC policies in action. The signalling flow is shown in Figure 3.7. We describe the signalling flow below.

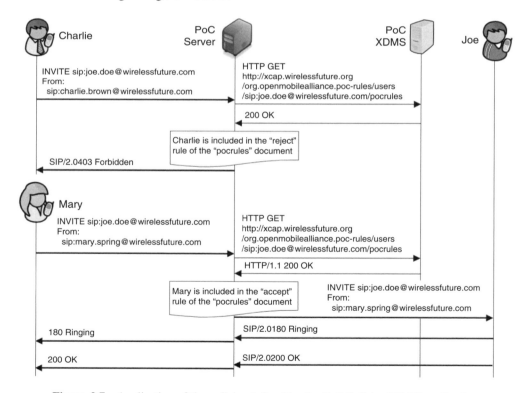

Figure 3.7   Application of the policies defined by the PoC Policies XCAP application

The scenario assumes that Joe (`sip:joe.doe@wirelessfuture.com`) has defined a PoC User Access policy document with a '`reject`' and an '`accept`' list. The former includes Charlie (`sip:charlie.brown@wirelessfuture.com`) and the latter includes Mary (`sip:mary.spring@wirelessfuture.com`). At this stage, the following actions occur:

1. Charlie intends to start an ad-hoc PoC session with Joe. Eventually, the request reaches the PoC server serving Joe (i.e. the Participating PoC server, although the split between Controlling and Participating functions is not shown in Figure 3.7 for sake of simplicity).
2. The PoC server contacts the PoC XDMS via POC-8 reference point. It retrieves the PoC rules document associated with Joe, using the well-known XCAP URI required for this purpose.
3. The PoC server finds Charlie's PoC identity included in the '`reject`' list, so it directly sends a `403 Forbidden` answer back to Charlie's PoC device, without sending any notification to Joe.
4. Later on, Mary tries to contact Joe as well. The server retrieves the PoC rules document again and finds out that Mary is included in the '`accept`' rule (NOTE).
5. Therefore, the server proxies the SIP INVITE request towards Joe's PoC client. The PoC answer mode configured at the time is '`manual`', so the regular Manual Answer Mode PoC signalling flow is executed in processing this request.
6. Eventually, Joe accepts the incoming call and the PoC session starts (further SIP, RTP and floor control messages are not displayed).

NOTE: Observe that the PoC server retrieves the PoC rules XML document each time a user tries to contact Joe.

3.6.5 XCAP Applications Supported by the Presence Enabler (Presence XDMS and RLS XDMS)

The OMA Presence enabler supports three additional XCAP applications, namely:

- The *Presence Subscription List* [11]. Having subscription lists enabled lets a watcher perform a single SIP transaction to subscribe to the Presence status of a potentially large number of users (instead of performing an individual subscription to each contact in the list). The main goal of this application is to save bandwidth in the wireless interface.
- The *Presence Subscription Authorisation Rules* [10]. This application defines, for a given user, which watchers are allowed to get Presence information from the Presentity and which are not.
- The *Presence Content Rules* [10]. This application defines what type of Presence information (e.g. location, availability, willingness, mood) can be provided to authorized watchers.

As one would expect, Subscription Authorization Rules and Presence Content Rules are built as a specialization of the Common Policy framework [21]. They are implemented by the

Presence XDMS. Subscription Lists are stored in the RLS XDMS, and they are a particular case of *resource lists* defined in [7].

Details about these three applications are provided in Chapter 4.

3.6.6 Summary of XCAP Applications

In this section we provide a general overview of the distribution of XCAP applications among the different XDM servers, as shown in Figure 3.8.

All XCAP servers must support the default XCAP Capabilities application. Additionally, XDM servers must support the XML Document Directory application which lets the client keep track of all documents stored in the XDM architecture. Another common functionality that all XDMSs must implement is user authorization.

The Aggregation Proxy appears as a single XDM server towards the XDMC, and is responsible for client authentication and traffic aggregation. It may perform data compression mechanisms as well.

Apart from the common functions supported by all XDMSs, OMA defines the applications which are specific to each OMA enabler (e.g. PoC, Presence), and must be implemented by enabler-specific XDMSs, such as the PoC XDMS, the Presence XDMS and the RLS XDMS. A Shared XDMS is also described, to support XCAP functions that may be shared across different OMA enablers. These applications have been described in detail through section 3.6. A graphical overview is shown in Figure 3.8.

As explained in section 3.4.4, client-server communication is based on the exchange of XML documents or sections of XML documents. In order to ensure that application policies and proper data validation are enforced by XCAP servers, it is mandatory to use application-specific MIME-Types (as opposed to using the general 'application/xml' type). This also applies to components of an XCAP document. Table 3.5 provides a summary of the valid MIME-Types that must be supported in all XCAP transactions. They are inserted either by

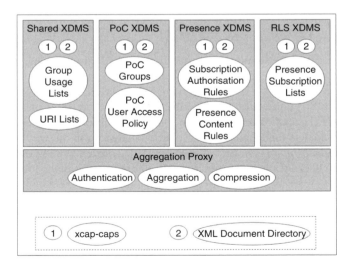

Figure 3.8 Split of functionalities across the elements of the XDM architecture

Table 3.5 Defined MIME-Types used by XCAP which are in scope of OMA XDM

Application	XDM Server	AUID	MIME-Type
XCAP Server Capabilities	All	xcap-caps	application/xcap-caps + xml
XML Documents Directory	All	org.openmobilealliance. xcap-directory	application/oma-directory + xml
URI List	Shared XDMS	resource-list	application/resource-lists + xml
Group Usage List	Shared XDMS	org.openmobilealliance. group-usage-list	application/vnd.oma.group-usage-list + xml
PoC Group	PoC XDMS	org.openmobilealliance. poc-rules	application/vnd.oma.poc.groups + xml
PoC User Access Policy	PoC XDMS	org.openmobilealliance. poc-rules	application/auth-policy + xml
Subscription List	RLS XDMS	rls-services	application/rls-services + xml
Subscription Authorization Rules	Presence XDMS	org.openmobilealliance. pres-rules	application/auth-policy + xml
Presence Content Rules	Presence XDMS	org.openmobilealliance. pres-rules	application/auth-policy + xml
XML Element	All	All	application/xcap-el + xml
XML Attribute	All	All	application/xcap-att + xml
XML Namespace definitions	All	All	application/xcap-ns + xml
Conflict report	All	All	xcap-error + xml

clients (when PUTting XML components) or by servers (when clients GET components from an XDMS), and must be supported by both.

It is worth observing that XCAP and its extensions defined by OMA XDM let any XDM client follow a clear process that will let it access all XDM documents and settings, without having to store a 'local' copy of all contents that are distributed across several XDMSs. Figure 3.9 illustrates this idea.

Observe that an XDMC device properly provisioned with a minimum set of information (i.e. XCAP identity, authentication credentials and XCAP root URI) may sequentially access all XML documents it needs to support any required user operation. For example, the XDMC may perform the following operations:

1. Upon client startup, the XDMC initially queries the XDM service (i.e. the Aggregation Proxy) about the supported XCAP capabilities. The Aggregation Proxy queries each XDMS and delivers an aggregated response to the client (the *back-end* queries are not shown in Figure 3.9).

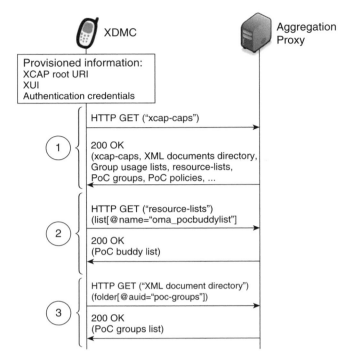

Figure 3.9   XCAP client initialization signalling flow

2. The XDMC retrieves the PoC buddy list, by downloading the URI List stored in the Shared
   XDMS with name 'oma_pocbuddylist' (refer to section 3.6.3.1).
3. Finally, the XDMC may invoke the 'XML Documents Directory' XCAP application to
   retrieve the list of PoC groups stored in the PoC XDMS. The XDMC may decide to
   download each Poc group XML document, or simply keep the group list and retrieve each
   group document whenever the user intends to initiate a group communication.

The signalling flow presented above is just one example of XCAP operations that can be
performed in an ordered way by the XDMC. Regardless of the actual sequence implemented
by a particular XDMC, the important aspect to remark is the fact that any properly configured
XDMC is able to manage all XML documents stored in the network in a consistent and secure
way (e.g. without the risk of losing information or producing unaccessible documents).

3.7 Summary and Conclusions

In this chapter, we have explained how the XDM enabler provides a common, reusable
mechanism to let users manage all XML configuration documents associated to any potential
SIP service. XML documents are used to define service configurations and policies, and they
fulfill functions as important as definition of PoC groups, implementation of policies and
blacklists, and creation of lists to be shared across applications.

We have presented the OMA XDM architecture as a realization of the more general XCAP framework defined by IETF. In particular, XCAP is a key element for any SIP based service requiring definition of per-user policies and settings. Therefore, we expect that new applications and capabilities will appear as new SIP services are standardized by OMA.

3.8 References

[1] OMA Work Item Document 'Group Management'; WID-0086, January 2004 (http://www.openmobilealliance. org/ftp/Public_documents/TP/permanent_documents/OMA-WID_0086-Group_Mgmt-V1_0-20040120-A.zip)
[2] J. Rosenberg: 'The Extensible Markup Language (XML) Configuration Access Protocol (XCAP)'; May 2007, RFC 4825.
[3] OMA XML Document Management (XDMv1.0.1): 'XML Document Management Architecture'; November 2006.
[4] OMA XML Document Management (XDMv1.0): 'XML Document Management Core Specification'; June 2006.
[5] OMA Push-to-Talk over Cellular (PoCv1.0): 'Push-to-Talk over Cellular (PoC) Architecture'; June 2006.
[6] OMA XML Document Management (XDMv1.0.1): 'Shared XDM Specification'; November 2006.
[7] J. Rosenberg: 'Extensible Markup Language (XML) Formats for Representing Resource Lists', May 2007, RFC 4826.
[8] OMA Push-to-Talk over Cellular (PoCv1.0.1): 'Push-to-Talk over Cellular (PoC) XDMS'; November 2006.
[9] OMA Presence (Presence v1.0.1): 'Presence SIMPLE Architecture'; November 2006.
[10] OMA Presence (Presence v1.0.1): 'Presence XDM Specification'; November 2006.
[11] OMA Presence (Presence v1.0.1): 'Resource List Server (RLS) XDM Specification'; November 2006.
[12] R. Fielding: 'Hypertext Transfer Protocol – HTTP/1.1', June 1999, RFC 2616.
[13] E. Rescorla: 'HTTP over TLS', May 200, RFC 2818.
[14] OMA Presence (Presence v1.0.1): 'Presence SIMPLE Specification'; November 2006.
[15] H. Sinnreich, A., B. Johnston, R., J. Sparks: 'SIP Beyond VoIP', VON Publishing LLC, New York, October 2005.
[16] T. Berners-Lee, R. Fielding, L. Masiner: 'Uniform Resouce Identifier (URI): Generic Syntax', January 2005, RFC 3986.
[17] J. Franks, et. al.: 'HTTP Authentication: Basic and Digest Access Authentication', June 1999, RFC 2617.
[18] 3GPP TS 33.222v7.2.0: 'Generic Authentication Architecture (GAA): Access to network application functions using Hypertext Transfer Protocol over Transport Layer Security (HTTPS) (Release-7)', October 2006.
[19] 3GPP TS 33.141v7.1.0: 'Presence service; Security (Release-7)', June 2006.
[20] J. Rosenberg, H. Schulzrinne, O. Levin: 'A Session Initiation Protocol (SIP) Event Package for Conference State', August 2006, RFC 4575.
[21] H. Schulzrinne, H. Tschofenig, J. Morris, et. al.: 'Common Policy: A Document Format for Expressing Privacy Preferences', February 2007, RFC 4745.
[22] J. Rosenberg: 'Presence Authorization Rules', Internet Draft (work in progress), IETF, October 2006.
[23] A., B. Roach: 'Session Initiation Protocol (SIP)-Specific Event Notification', June 2002, RFC 3265.
[24] OMA XML Document Management (XDMv1.1): 'XML Document Management Architecture'; August 2007 (Work in Progress).
[25] J. Urpalainen: 'An Extensible Markup Language (XML) Configuration Access Protocol (XCAP) Diff Event Package', August 2007 (Work in Progress).

4

The OMA Presence Service

4.1 Introduction

The third enabler defined by the Open Mobile Alliance within the scope of the OMA PoC and PAG Working Groups is *OMA SIMPLE Presence*. The Presence service provides users with dynamic information such as status and availability of their contacts. Presence information lets the user decide which is the most effective and non-intrusive way to communicate with each of her contacts. Consequently, the Presence service increases the likelihood for success of the user's communications, and enhances user satisfaction.

For operators and service providers, Presence may represent a new revenue stream – users may be charged to receive Presence updates about their contacts – but, more importantly, it lets them provide a new user experience to their customer base. As we will see, the Presence service may evolve to deliver to users a *rendevouz point* for all their communication means, in a user-friendly, secure and non-intrusive way. Hence, the Presence service and its evolution may significantly change the way users perceive and interact with the IP multimedia communication technologies that will be at their hand. The consequence of this shift in how users perceive their communications will obviously change the way we interact with our friends, family or colleagues.

Presence is a concept much broader than the OMA Presence enabler. Intense research and standardization activities have focused on Presence during the last ten years, particularly in the IETF and 3GPP. In addition, many implementations of Presence services have been available on the Internet for a while, supporting applications such as gaming, chat, messaging or VoIP. Covering all aspects and implications of the Presence service falls out of the scope of the present volume. The goal of the present chapter is to provide the reader with enough detailed information about the Presence concept and OMA Presence in particular, as a basis to understand how Presence and PoC can interoperate and provide a consistent user experience to the mobile user.

The IETF has been very active in developing a SIP-based framework to support a standards-based Presence service. In particular, the IETF SIMPLE Working Group was created to extend core SIP to provide support for Instant Messaging and Presence (SIMPLE stands for *SIP for Instant Messaging and Presence Leveraging Extensions*).

Multimedia Group Communication. Andrew Rebeiro-Hargrave and David Viamonte Solé
© 2008 John Wiley & Sons, Ltd.

The first attempt to define a standard Presence service tailored for wireless scenarios was carried out in the *Wireless Village* (WV) consortium. OMA inherited WV specifications and developed the *Instant Messaging and Presence service* (IMPS) enabler [1]. IMPS, however, is not based on the broad Presence framework defined by IETF. In order to avoid fragmentation and to ensure interoperability of Internet and mobile Presence services, OMA initiated the definition of a SIMPLE-based Presence service. Such activity was formalized in Work Item WI0073 [2].

Within the 3GPP/3GPP2 framework, the initiative to standadize a SIMPLE-based Presence service over IMS was also undertaken within the Release-6 timeframe. In order to avoid overlapping of activities, 3GPP, 3GPP2 and OMA eventually agreed a work split on Presence, which would leave OMA responsible for further definition of a mobile capable Presence service. 3GPP/3GPP2 would focus on the integration of Presence capabilities into their IMS framework [3, 4].

In a nutshell, the goals of the OMA Presence Working Group are:

* To specify a fully standard SIP/SIMPLE Presence service, by further consolidating and developing already existing IETF/3GPP work with the necessary extensions required to guarantee service implementability and interoperability;
* To develop a Presence standard which should support mobile users in an efficient way (i.e. without wasting network resources in the wireless domain);
* To develop a service that should take OMA PoC as its first *customer* to be enriched with Presence capabilities.

In the particular case of PoC, Presence represents a powerful tool: for instance, a user may be interested in knowing who are the available contacts for an ad-hoc group call. For one-to-one calls, a user may use Presence information about the *to-be-contacted* user to decide among: a) starting a PTT session immediately (if the user is available), or: b) sending a call-back request (Instant Personal Alert), if the user is attending a meeting: all these possibilities are enabled by deploying a Presence service to support PoC subscribers.

If we think about a user accessing her PoC contact list in the User Interface of her PoC device, the typical way in which Presence information enriches user experience consists of using mnemonic icons that display Presence information at a glance. Figure 4.1 provides an

Figure 4.1   Example PoC contact list enhanced with Presence information

example, where the user may see whether her contacts are *online* (e.g. Bert), *offline* (e.g. Ally), or who has enabled the PoC automatic answer mode (e.g. Adam).

The above example shows a very simple use case that illustrates how Presence information can be presented to the PoC user and how he can use it. Obviously, there are much more complex use cases and ways of using Presence information.

The next section will provide a description of general Presence concepts, while the rest of the chapter will focus on the OMA Presence Enabler in particular.

4.2 General Presence Concepts

4.2.1 Entities Involved in the End-to-End Presence Service

At first sight, we can think Presence as a service that lets users gain awareness of status information about some of their contacts. Status information may include *availability*, *terminal capabilities*, *willingness to engage in communication* (e.g. a PoC session), *location . . .*

In addition to human users, Presence information can also be generated by and provided from/to non-human applications. For example, a SIP-based application may subscribe to receive Presence updates about a certain user, and thus customize the service offering based on the status of that user at a given time. Consider, for example, a server delivering adverts to cellular subscribers: such an application may personalize each advert based on dynamic Presence information of the potential recipients.

Example non-human Presentities could be: a radio broadcast station (*I wish to subscribe to Presence information about my favourite programs*) or a domotic home (*I wish to receive a Notification when I am away and a door or window gets opened*).

Figure 4.2 presents a general overview of the different elements involved in the end-to-end Presence service. We briefly describe them below.[1]

The *Presentity* (contraction for *Presence Entity*) is a logical element that has Presence information associated with it. A Presentity is an addressable resource which generally represents a person, but it may also represent any other resource or service. A Presentity can be associated to a helpdesk application (Presence can be associated to the availability of at least one operator), a meeting facility (Presence information used to manage bookings) or any other resource whose associated information can be shared via Presence services. The address of a Presentity is in the form of a SIP URI or a TEL URI.[2]

The Presence architecture makes a clear distinction between the function performed by the Presentity (having Presence information associated with it) and the function performed by the Presence Source. We can think the Presence Source as a logical function that communicates with the network *on behalf of* a single Presentity, and provides Presence information related to that Presentity.

---

[1] There is extensive literature through IETF (e.g. [5]), 3GPP and OMA about the different entities involved in the end-to-end propagation of Presence information. In fact, there are several minor misalignments between IETF and OMA in terms of defining some Presence-related concepts. When differences exist, we will follow the OMA definition for a given term, which is generally aligned with the 3GPP concept as well.

[2] The 'pres:' URI has also been defined to address Presentities [6]. However, 'sip:' and 'tel:' schemas are expected to be more relevant for Presence services, since they allow a direct relationship between SIP-based services and the Presence enabler.

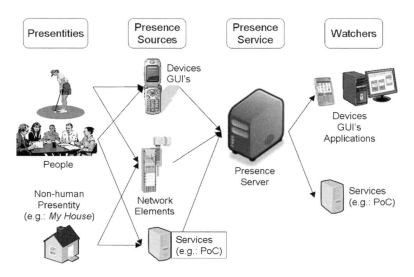

Figure 4.2   Elements involved in the end-to-end Presence service

The standard mechanism used by Presence Sources to update Presence information is sending a SIP PUBLISH message towards the Presence network entity. This request contains a special body that indicates the updated Presence parameters. Hence, we can say that Presence Sources *publish* Presence information.

For example, when Ron logs on into a SIP/IP Core and becomes available, the end user device contains a SIP/SIMPLE Presence client that behaves as a Presence Source. The client communicates the new status to the Presence Service, while sip:ron@example.com is the Presentity on which behalf the Presence Source sends Presence publications.

The split of functions between the Presentity and the Presence Source makes it possible for several Presence Sources to publish Presence information related to the same Presentity. Observe that Presence information may cover a broad range of topics, such as registration status, terminal capabilities, availability, willingness to communicate, contact address(es), location, and images. In general, all this information may reside in several elements distributed between the client and the network. Consequently, the Presence architecture supports that different entities behave as a Presence Source for the same Presentity. This approach lets the Presence Service build a complete set of Presence information from the different contributions received from each source.

This approach, however, is not always necessary: in some cases, only one or two Presence Sources may provide a reduced but complete set of information, which covers the requirements of most customer needs. In the particular case of a basic PoC service, knowing whether a user is online (i.e. registered with IMS) and if he/she has a PoC-enabled device, provides enough Presence information to all potential watchers. On the other hand, advanced PoC services that include support for emergency services and crisis management would require additional Presence information such as user location.

We have learned that it is possible that a single Presentity has several Presence Sources publishing complementary Presence information on behalf of it. Similarly, it is feasible for a single network element to behave as a Presence Source for several Presentities. In any case, each publication of Presence information performed by a Presence Source must be associated to one and only one Presentity. Figure 4.3 shows an example of how two Presence Sources publish Presence information on behalf of the same Presentity: in this case, the user's terminal provides basic Presence information, while a Location Server supplies location information about that user. The Presence Service aggregates this information to build a single Presence information document with all received information.

We have seen that the Presentity and the Presence Source are the two elements involved in providing Presence information updates towards the Presence Service. The entities that are *customers* of this information are the *Watchers*: a watcher is any entity that requests reception of Presence information about one or more Presentities. As we will see below, in order for a watcher to receive Presence information, it must subscribe to receive Presence updates about the Presentities; this is achieved by means of the SIP SUBSCRIBE message.

The *Presence Server* is the logical entity located at the network of the service provider. The Presence server receives Presence information updates from Presence Sources. For each Presentity, the Presence service propagates Presence information towards all watchers subscribed to Presence information about that Presentity.

The *Presence Server* logical function performed by the network is generally implemented through a combination of different entities. For this reason, we will differenciate between the more general *Presence service*, as the set of logical functions supporting the end-to-end Presence service, and the *Presence server*, as the SIP Application Server responsible for handling subscriptions, publications and notifications. We will further describe how the realization of the Presence service requires several different logical functions in section 4.3.

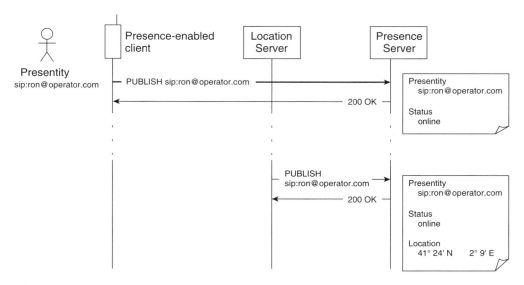

Figure 4.3 Two Presence Sources publishing Presence information on behalf of a common Presentity

We have presented the responsibility of the Presence service merely as a relay function of Presence information between Presence Sources and watchers. However, this is a simplistic view, which does not cover other key functions such as Presence authorization or content filtering. We will present them in section 4.3, in the framework of the OMA Presence service.

4.2.2 Sample Signalling Flow

Putting the pieces presented above together, we can see that the Presence server receives *Presence Publications* (i.e. SIP PUBLISH messages) from each Presence Source, and *Presence subscriptions* (i.e. SIP SUBSCRIBE messages) from watchers asking reception of Presence updates. Once a watcher is correctly subscribed to receive Presence information about one or more Presentities, the Presence service notifies the watcher each time a Presence update is received. The server uses the SIP NOTIFY message to send updated Presence information towards a watcher. This is summarized in the figure below.

Figure 4.4 shows a typical Presence signalling flow, where Presence Sources update the status of the Presentity and watchers subscribe to the service and receive a notification every time the status of one Presentity changes:

1. A Presence Source uses SIP PUBLISH to inform the Presence Server about the status of the Presentity (sip:ron@wirelessfuture.com). The Source includes an Event: presence header that indicates that it will publish Presence information. Given that the publication-subscribe mechanism is generic; this header indicates which application package motivates this SIP PUBLISH message (as opposed to other SIP applications that use the SIP PUBLISH method for other purposes).

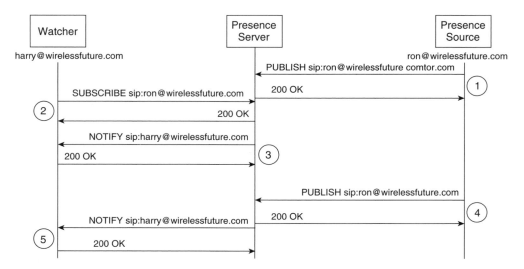

Figure 4.4   Typical Presence signalling flow

```
PUBLISH sip:ron@wirelessfuture.com SIP/2.0
Via: SIP/2.0/UDP ronhost.wirelessfuture.com;branch=
z9hG4bK652hsge
To: <sip:ron@wirelessfuture.com>
From: <sip:ron@wirelessfuture.com>;tag=1234wxyz
Call-ID: 81818181@ronhost.example.com
CSeq: 1 PUBLISH
Max-Forwards: 70
Expires: 3600
Event: presence
Content-Type: application/pidf+xml
Content-Length: . . .


(Presence PIDF document)
```

Observe that the Presence publication is valid for one hour (value of the `Expires:` header).

2. A watcher (`sip:harry@wirelessfuture.com`) installs a subscription in the Presence server requesting information about Ron. The watcher also includes an `Event: presence` header to indicate that this subscription requests Presence information about this Presentity.

```
SUBSCRIBE sip:ron@wirelessfuture.com SIP/2.0
Via: SIP/2.0/TCP harryhost.wirelessfuture.com;branch=
  z9hG4bKnashds7
To: <sip:ron@wirelessfuture.com>
From: <sip:harry@wirelessfuture.com>;tag=xfg9
Call-ID: 2010@harryhost.wirelessfuture.com
CSeq: 17766 SUBSCRIBE
Max-Forwards: 70
Event: presence
Accept: application/pidf+xml
Contact: <sip:harry@wirelessfuture.com>
Expires: 7200
Content-Length: 0
```

The server accepts this subscription (200 OK) and indicates its duration (which may be equal or lower than 7200 seconds). Before the expiration time is reached, the watcher may reissue a SUBSCRIBE request to renew the subscription.

3. The server immediately sends a SIP NOTIFY message informing about the latest known status of the Presentity. The contents of the NOTIFY document is an XML encoded Presence document, as defined in [11]. We describe this format in section 4.2.3. In case the server does not have any information about the Presentity, it may also send an empty SIP NOTIFY message.

```
NOTIFY sip:harry@wirelessfuture.com SIP/2.0
Via: SIP/2.0/UDP presence.wirelessfuture.com;branch=
   z9hG4bK8sdf2
To: <sip:harry@wirelessfuture.com>;tag=12341234
From: <sip:ron@wirelessfuture.com>;tag=abcd1234
Call-ID: 12345678@presencehost.wirelessfuture.com
CSeq: 1 NOTIFY
Max-Forwards: 70
Event: presence
Subscription-State: active; expires=3599
Contact: sip:presence.example.com
Content-Type: application/pidf+xml
Content-Length: . . .

(Presence PIDF document)
```

4. After a while, the status of a Presentity changes (e.g. user `sip:ron@wirelessfuture.com` joins a meeting and turns his status into *busy*). The Presence Source PUBLISHes the new Presence status towards the Presence server.
5. The Presence server NOTIFies the watcher about the new status of the Presentity.

Observe that Presence signalling flows follow standard SIP format, structure and routing rules as per [7]. In addition, the IETF SIMPLE WG has defined three main extensions to define how Presence information can be exchanged over SIP messages, namely:

- [8] defines the SIP *event notification framework*: a generic mechanism to let SIP clients subscribe to certain events, and receive notifications whenever the subscribed event generates some information. This framework is based on the two new SUBSCRIBE and NOTIFY methods defined in [8] as well.
- [9] specifies the so-called *Presence event package*: it defines how to use the general event notification framework [8] to handle Presence subscriptions and notifications.
- [10] defines the mechanism used by Presence Sources to publish Presence information into a Presence Server. The main method defined in [9] is SIP PUBLISH.

The signalling flows presented in this section follow the mechanisms defined in these three references. Additionally, as explained above, another key building block required to build up the SIP Presence framework is the definition of the actual Presence information format exchanged in NOTIFY and PUBLISH messages. Such format is defined in [11].[3]

---

[3] Following IETF's modular approach, the Presence information format is independent from the carrying protocol (e.g. SIP). This allows for greater flexibility and independency: theoretically, SIP can carry any well-defined Presence information format apart from [11] and, in turn, Presence information can be carried over other protocols apart from SIP (e.g. HTTP).

4.2.3 Formatting Presence Information: the Presence Information Data Format

We have seen how the Presence service defines a set of roles and how three basic SIP messages (PUBLISH, SUBSCRIBE, NOTIFY) are used to propagate Presence information between sources and watchers. We still need to explain *what* Presence information *is*: what are the contents carried in PUBLISH and NOTIFY messages? We devote this section to answer this question.

IETF defined the XML-based *Presence Information Data Format* (PIDF) [11] as the basis to support communication of general Presence information across two entities. PIDF was developed within the IETF *Instant Messaging and Presence Protocol* (IMPP) Working Group. The idea was first to create a consistent set of basic requirements which should be met by any application supporting exchange of Presence information [12]. PIDF was then developed as the extensible format that should meet the original Presence requirements from [12, 13].

PIDF is a transport-protocol agnostic format. Therefore, SIP/SIMPLE uses SIP as the framework to support publication and notification mechanisms, while actual information is carried in the PIDF body of SIP messages. For that purpose, the `application/pidf+xml` MIME type has also been defined.

A simple PIDF document is shown below:

```xml
<?xml version="1.0" encoding="UTF-8"?>
<presence xmlns="urn:ietf:params:xml:ns:pidf"
    entity="sip:ron@wirelessfuture.com">
  <tuple id="sr6tree79">
    <status>
       <basic>open</basic>
    </status>
    <contact priority="0.8">sip:ron@wirelessfuture.com
    </contact>
    <timestamp>2007-03-22T10:25:01Z</timestamp>
  </tuple>
  <note>I'll be in a concert tomorrow evening</note>
</presence>
```

PIDF documents are associated with the XML namespace 'urn:ietf:params:xml:ns:pidf'. The structure of the PIDF document is as follows:

- Each PIDF document must contain a root `<presence>` element declared under the PIDF namespace defined by [11]. This element must contain an 'entity' attribute, which contains the address (e.g. SIP URI) of the Presentity that motivates the publication of the document. When other namespaces are used to extend the basic PIDF functionality, they must also be declared in the root element. The `<presence>` element may contain zero or more `<tuple>` elements, followed by zero or more `<note>` elements. Optional extension from other namespaces may follow.

- The `<tuple>` element contains a set of Presence information that generally has a certain meaning on its own. For example, it may contain the status associated to a given service (e.g. PoC). The `<tuple>` element contains:
  - ○ a mandatory `<status>` child element, followed by the optional elements:
  - ○ zero or more extension elements (possibly from other namespaces), followed by:
  - ○ zero or one `<contact>` element, followed by:
  - ○ zero or more `<note>` elements, followed by:
  - ○ zero or one `<timestamp>` element.
- The `<status>` element must contain at least one child element. The default element defined by PIDF is the optional `<basic>` element, whose defined values are 'open' and 'closed'. When present, the `<basic>` element generally indicates communication availability of the Presentity for a certain application. In case of OMA PoC, this element indicates user availability to receive PoC session invitations. The exact meaning of the `<basic>` values is application dependant, but we can associate 'open' to 'available', and 'closed' means 'unavailable'.
- The `<contact>` element provides a contact address that may be used by watchers to contact the Presentity. For example, this may contain SIP, TEL or mailto: contact details.
- The `<note>` element may be used to insert human-readable comments. `<note>` elements may be related to tuples or to the whole `<presence>` element, depending on their location within the PIDF document. In any case, `<note>` elements cannot be used to override the value of the `<status>` value provided in a given `<tuple>` element.
- The `<timestamp>` element indicates the moment when current information about the `<tuple>` was published.

Considering all this information, we can now understand the message shown above. The document refers to Presentity `sip:ron@wirelessfuture.com`, and it contains a single `<tuple>` element. It indicates that Ron is available for a generic service (the document does not provide additional details about it). A contact address is provided (it happens to be the same as the Presentity URL). We can see that `<tuple id='sr6tree79'>` was last updated on March 22nd, 2007.

The document provides a final `<note>` element containing human-readable text.

4.2.4 Extending PIDF

At this stage, we understand the basic format of PIDF documents. This provides a general and extensible framework to support exchange of simple Presence information. It is clear, however, that the set of information defined by PIDF is very small. Although it can be extended by the user, doing it outside the umbrella of standardization bodies such as IETF or OMA would lead to interpretation and interoperability issues.

Having this idea in mind, the industry soon noticed that there was a lot of potential in providing a richer set of well-defined (and, still, extensible) Presence information. This could include not only communication availability, but also the user's willingness and mood, device status and capabilities, location, and contact information.

Hence, a number of extensions to plain PIDF have been developed, and still more are being developed. We briefly describe them in the next sections.

4.2.4.1 The Presence Data Model

[11] defines a Presence document as a series of tuples that contain status information, several optional fields and additional optional user-defined markup. PIDF does not clearly define what a tuple is meant to be, and this makes it difficult to map information obtained from real life systems and Presentities into well-structured and interoperable Presence information.

The *Presence Data Model* [14] defines the different components that contribute to build up a complete set of Presence information. This model is outlined in Figure 4.5.

The goal of the Presence Data Model is to build up a coherent and structured description of a Presentity (referenced by a URI). In order to achieve this goal, three components of the model are defined: the *Person*, the *Service(s)* and the *Device(s)*.

Each Presence document refers to a single *Person* component. However, given the inability of a Presence service to determine what a real person is, this concept can be extended to model human and non-human entities (e.g. a helpdesk service, a news feeder, a weather site). Regardless of this generalization, the Presence document must refer to a unique *Person* entity.

Each person is in general capable of using several devices to access zero or more services. There may be services which are executed in one device (e.g. a PoC handset), devices supporting several services (e.g. a computer supporting mail, IM, VoIP and videotelephony) and devices supporting a single application (e.g. a pager). Consequently, the Presence Data Model supports a 'many-to-many' relationship between services and devices, under the umbrella of a single Presentity.

Most Presence information we can think of can be classified into these three categories. Effectively, a watcher may be interested in:

- Knowing information about location, current activity and willingness to communicate of a *person*.
- Having information about the *services* that a person is able to use and the availability for communication in each of those services.
- Knowing which type of *devices* the user is using/able to use, to determine which is the most effective type of communication to set up a successful session. For example: if a user

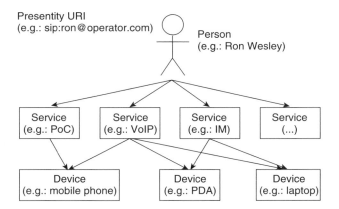

Figure 4.5   The Presence Data Model

is engaged in a conversation using his mobile phone, it is unlikely that he will be able to answer an incoming call, but he will possibly read a short message after finishing the current session.

In order to map this model into the Presence document format, [14] defines two new XML elements, namely: the `<person>` and the `<device>` elements. These are used to convey information about the person and the device respectively.

Information about services is included in the `<tuple>` element already defined by [11]. Each service must be associated to one tuple, so if a user has Presence associated to two services, two `<tuple>` elements must be used in the document. Each tuple has a different 'id' value associated. The `<contact>` sub-element, if present inside a `<tuple>` element, provides a URI suitable to contact the Presentity using the service associated to that tuple.

Finally, the Presence Data Model proposes a distinction between temporary (status) information, which may change over the time, and permanent (characteristics) information about a person, device or service.

We can see below a sample Presence document that implements the Presence Data Model.

```
<?xml version="1.0" encoding="UTF-8"?>
<presence xmlns="urn:ietf:params:xml:ns:pidf"
    xmlns:dm="urn:ietf:params:xml:ns:pidf:data-model"
    xmlns:rp="urn:ietf:params:xml:ns:pidf:rpid"
    xmlns:xsi=http://www.w3.org/2001/XMLSchema-instance
    entity="sip:ron@wirelessfuture.com">

 <tuple id="sg89ae">
   <status>
    <basic>open</basic>
   </status>
   <dm:deviceID>mac:8asd7d7d70</dm:deviceID>
   <contact>sip:ron@pc122.wirelessfuture.com</contact>
 </tuple>

 <dm:person id="p1">
   <rp:activities>
    <rp:meeting/>
   </rp:activities>
 </dm:person>

 <dm:device id="pc122">
   <rp:user-input>idle</rp:user-input>
   <dm:deviceID>mac:8asd7d7d70</dm:deviceID>
 </dm:device>
</presence>
```

The PIDF, Presence Data Model and *Rich Presence Information Data* (RPID) [15] namespaces are declared at the beginning of the document.

The document contains a `<person>` element that includes an `<activities>` child element (Ron is currently in a meeting). A `<device>` element provides information about the status of a device (it is *idle*: the user is not active with that device when publishing this information) and one permanent characteristic, such as the `<deviceID>` (which in this case is based on a MAC address).

4.2.4.2 Rich Presence Information Data Format (RPID)

The *Rich Presence extensions to the Presence Information Data Format* (RPID) [15] have been developed in parallel with [14]. If the Presence Data Model provides a coherent way of organizing Presence information, RPID defines the extensions that make Presence information a much richer and powerful service when compared with the support offered by plain PIDF.

RPID defines a new set of elements and attributes that can be included in Presence documents to enrich, enhance and detail Presence information about a Presentity. All elements defined by RPID are aligned with the Presence Data Model, so they should be placed inside the `<tuple>`, `<person>` or `<device>` elements of the Presence document.

The following example shows how a `<tuple>` and a `<person>` element can be enriched with new RPID-defined elements, such as status icons (the Presentity may provide external links to images that provide an iconic representation of some Presence information), person's mood (e.g. `<happy>`), activities the person is involved in (e.g. `<vacation>`) and expected time boundaries of these activities. For a complete list of elements defined by RPID, the reader is referred to [15].

```
<tuple id="eg92n8">
   <status>
     <basic>open</basic>
   </status>
   <dm:deviceID>urn:x-mac:0003ba4811e3</dm:deviceID>
   <rpid:class>email</rpid:class>
   <rpid:service-class><rpid:electronic/></rpid:service-
   class>
   <rpid:status-icon>http://example.com/mail.png</rpid:
   status-icon>
   <contact priority="1.0">mailto:ron.wesley@hogwarts.com
   </contact>
</tuple>

<dm:person id="p1">
   <rpid:activities from="2006-05-30T12:00:00+05:00"
     until="2006-06-15T17:00:00+05:00">
     <rpid:note>Far away</rpid:note>
```

```
      <rpid:vacation/>
    </rpid:activities>
    <rpid:mood>
      <rpid:happy/>
    </rpid:mood>
    <rpid:place-is>
      <rpid:audio>
        <rpid:noisy/>
      </rpid:audio>
    </rpid:place-is>
    <rpid:place-type><rpid:residence/></rpid:place-type>
    <rpid:status-icon>http://example.com/play.gif</rpid:
    status-icon>
  </dm:person>
```

4.2.4.3 Presence-based GEOPRIV Location Object Format

[16] defines an additional Presence extension that lets Presentities provide different types of location information. These may include regular addressing information (city, street, number) or GPS coordinates using the Geographic Markup Languange (GML) [17]. When this option is used, the GML namespace must be declared together with the rest of namespaces (e.g. PIDF, RPID, GEOPRIV).

Together with the inclusion of location information in Presence documents, [16] defines a simple set of rules that let users define who is able to receive location objects, which may be one of the most sensible information shared between Presentities and watchers.

Provided that privacy issues are correctly handled by the Presence service, the ability to provide location information in Presence documents opens a broad space of value added services. Consider the following example, where a messaging service subscribed to Presence information delivers a message to a user when it detects it has entered a certain area.

In the case depicted in Figure 4.6, the *Messaging Server* subscribes to receive Presence information about Ron (including Ron's location information):

1. The Location Server periodically updates (SIP PUBLISH) location information to the Presence Server (SIP Core functionality is not displayed for sake of simplicity).
2. For each Location update, the Presence Server sends a notification (SIP NOTIFY) to the Messaging Server.
3. The Messaging Server detects that Ron has entered a preconfigured 'trigger' area. It then sends a message to Ron (e.g. 'There is a famous restaurant close to your current location. You will find it at the following address . . .'), according to a pre-configured alert service that Ron has subscribed to.

Apart from the simple use case depicted above, other Value Added Services based on Location extensions to the Presence service can be deployed. There is currently ongoing investiga-

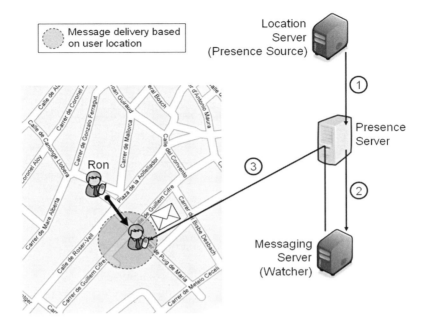

Figure 4.6   Location-based Presence use case

tion on how to provide location information in a secure and end user (i.e. Presentity) controlled way [18].

4.2.4.4 Other Presence Document Format extensions

The extensions described above (plain PIDF, RPID and the Presence Data Model) are reused by the OMA Presence v1 enabler. However, it is worth noting other Presence information extensions defined by IETF:

- *Contact Information in Presence Information Data Format (CIPID)* [19]. This extension lets a Presentity enrich her Presence document with contact information such as a business card, sound, homepage reference or an icon.
- *Timed Presence Extensions to PIDF* [20]. This extension lets a Presentity inform about events that happened in the recent past or those which are planned for the near future. As an example, a Presence Source application can use this extension to provide information about expected future status, location and activities, based on a calendar or agenda application.
- *Session Initiation Protocol (SIP) User Agent Capability Extension to Presence Information Data Format (PIDF)* [21]. This is a particularly interesting extension used to indicate SIP capabilities in Presence documents. As an example, it may provide information about supported media formats (e.g. audio, video, text), half-duplex vs. full duplex capabilities, available SIP methods and addressing schemes (e.g. `sip:`). This is useful to provide *hints* to Presence watchers in order to properly contact the Presentity using SIP-based services (e.g. PoC, IM, VoIP).

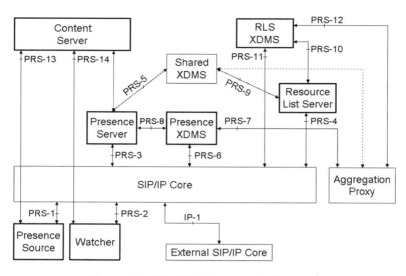

Figure 4.7 The OMA Presence Architecture[4]

4.3 The OMA Presence Service

4.3.1 Introduction and Architecture

In order to ensure market adoption, vendor competition and interoperability, the OMA Presence service is fully based on IETF standards. Hence, OMA Presence basically defines an implementation of the SIP/SIMPLE service specified by IETF, thus reusing concepts such as the generic subscription / notification mechanism, the Presence event package and the PIDF format and its extensions ([8, 9, 11]).

Figure 4.7 presents the OMA architecture for Presence [22]. The reader will identify several elements which are already familiar as they have been presented in previous chapters about PoC and XDM. Other Presence-specific nodes such as the Presence Server, the Watcher and the Presence Source have been briefly described in section 4.2. The main aspects of the OMA Presence architecture are described below.

PoC and XDM specific elements have been omitted from Figure 4.7 for sake of clarity. In general, however, the device used to access the PoC service may implement PoC client, XDM client, watcher and possibly Presence Source functionalities.

Functionality of the Presence Source (publication of Presence information on behalf of a Presentity) and the watcher (subscription to Presence information of a Presentity) have been presented in the previous section, and are not modified by the OMA Presence specification. OMA supports the following types of watchers:

---

[4]NOTE: At the time of writing (September 2007) OMA is about to approve the second revision of the Presence v1 enabler (OMA Presence v1.1). An important architectural change from the depicted picture is the removal of interfaces PRS-6 and PRS-11, due to the withdrawal of a mechanism to subscribe to changes in XML documents. Although interface numbering will be kept to ensure backwards compatibility with Presence v1.0 and v1.0.1 naming conventions, its functionality will finally not be included in OMA Presence v1.1 [35].

- *Subscribed Watchers* behave as described in section 4.2.2: a subscription is installed in the Presence server. For the duration of that subscription, the Presence server notifies Presence changes to the watcher. Shortly before the subscription expires, the Watcher may decide to renew the subscription by re-issuing a SIP SUBSCRIBE request.
- *Fetchers* are *one-shot watchers*: a SIP SUBSCRIBE message is sent to retrieve Presence information available at the moment the subscription is received. The SUBSCRIBE message contains an `Expires: 0` header, indicating that no subscription shall be stored at the Presence server, so that the fetcher is not notified of further Presence changes. This type of behaviour is useful for Presence clients that offer the user the possibility to perform an instant update of Presence information, but do not wish to handle a subscription mechanism.
- *Pollers* are periodic fetchers. In this case, a fetch message (SUBSCRIBE with Expires: 0 header) is sent periodically. This is a particular case of a fetcher client. This mechanism may not be very efficient if the polling frequency is high.

The behavioral difference between a regular watcher, a fetcher and a poller is outlined in Figure 4.8.

The main responsibilities of a Presence server have been introduced before (handling Presence publication and subscription requests, and issuing SIP NOTIFY messages accordingly). However, OMA provides a more detailed description of the tasks that a Presence server should perform when implementing them. We will briefly describe the OMA Presence server functionality in section 4.3.3.

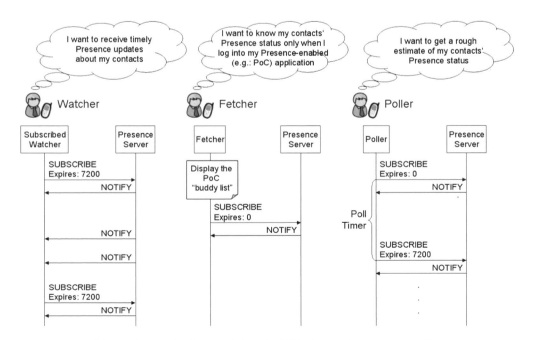

Figure 4.8 Behavior of a Subscribed Watcher, a Fetcher and a Poller

The *Resource List Server* (RLS) is a key element of the Presence architecture. The RLS is used to handle subscription to *Presence lists*. This functionality lets users (i.e. watchers) subscribe to receive Presence information about a list of Presentites, rather than sending an individual subscription for each Presentity. The goal is to achieve a substantial reduction of the traffic exchanged between the watcher and the network, which generally traverses a wireless (and possibly resource–limited) link. The RLS functionality is presented in further detail in section 4.4.

The *Content Server* provides storage of content such as audio files or images, which can be referenced – as we have presented in section 4.2.4.2 – in Presence documents. Watchers can download the referenced files to complement and enrich the Presence information about a Presentity. Presence Sources are responsible for uploading files to the content server and providing the right references in the Presence document attached to the PUBLISH message.

There are two XDMS elements associated to the Presence architecture, namely: the Presence XDMS and the RLS XDMS:

- The Presence XDMS is used to store *Presence authorization* documents, which determine *who* (which watchers) is authorized to receive *what* Presence information about a Presentity. The end user (i.e. the Presentity to which the document refers) uses this document to filter Presence information and prevent watchers from being able to receive sensitive information.
- The RLS XDMS manages *Presence subscription lists*. These documents are used by watchers to perform subscription to lists, rather than individual subscription to Presentities.

Both the Presence authorization list and the Presence subscription list documents may refer to lists of users. Therefore, external references to lists stored in the Shared XDMS can be used. In this case, reference points PRS-5 and PRS-9 can be used by the Presence server and the RLS respectively to retrieve the corresponding URI list document.

In addition to the elements defined by the OMA Presence architecture [22] there are three external elements required to support this service:

- The *SIP/IP Core* logical function provides support for user authentication and registration, service discovery, SIP message routing, signalling compression and subscriber provisioning. When the service is deployed in a 3G network, the default SIP/IP Core is the IMS as specified by 3GPP. As we will see in chapter 5, having IMS as the common SIP/IP Core for PoC, XDM and Presence lets a service provider combine these services and deliver a richer and more complete user experience to OMA subscribers. The presence of the `Event: presence` header in Presence-related SIP messages (e.g. SUBSCRIBE) is a key information used by the SIP/IP Core to route messages to the Presence server.
- The *Shared XDMS* provides storage of URI list documents that can be used as a basis to construct Presence subscription lists or Presence authorization documents.
- The *Aggregation Proxy* supports XDMC authentication and relay of XML documents between XDM clients and the different XDM servers, including the Presence XDMS and the RLS XDMS.

4.3.2 Presence Reference Points

We provide here a brief overview of all the OMA Presence reference points, together with their function and the protocol used (Table 4.1).

Since OMA PoC and OMA XDM reference points have been presented previously, we will not go into much detail about those OMA Presence reference points whose associated functionality can be directly derived from similar interfaces in PoC and/or XDM enablers. Although different SIP messages may be used in different enablers (e.g. SIP INVITE vs. SIP SUBSCRIBE) the core purpose of some interfaces can be well understood by simple comparison with PoC and XDM architectures. For example, PRS-1 interface connects a 'Presence

Table 4.1 OMA Presence reference points

Interface	Usage	Protocol
PRS-1	Publication of Presence information on behalf of a Presentity	SIP
PRS-2	Handling of watcher subscriptions (SUBSCRIBE and NOTIFY messages exchanged between the watcher and the SIP/IP Core)	SIP
PRS-3	Communication between the SIP/IP Core and the Presence server to relay SIP SUBSCRIBE, NOTIFY and PUBLISH transactions	SIP
PRS-4	Communication between the SIP/IP Core and the RLS to handle subscription to a list of Presentities	SIP
PRS-5	Communication between the Presence server and the Shared XDMS to retrieve URI lists. This is required when an external URI list is referenced in a Presence authorization rules document	XCAP
PRS-6 (withdrawn)	Communication between the Presence XDMS and the SIP/IP Core to serve subscriptions to changes in Presence authorization rules documents	SIP
PRS-7	Communication between the Aggregation Proxy and the Presence XDMS to support XML document management (e.g. creation, retrieval, modification, deletion) upon XDMC request	XCAP
PRS-8	Communication between the Presence Server and the Presence XDMS to retrieve Presence authorization rules XML documents	XCAP
PRS-9	Communication between the RLS and the Shared XDMS to retrieve URI lists. This is required when an external URI list is referenced in a Presence list	XCAP
PRS-10	Communication between the RLS and the RLS XDM to retrieve Presence lists	XCAP
PRS-11 (withdrawn)	Communication between the SIP/IP Core and the RLS XDMS to support subscription to changes in Presence list documents	SIP
PRS-12	Communication between the Aggregation Proxy and the RLS XDMS to support XML document management upon XDMC request	XCAP
PRS-13	Used by the Presence Source to upload MIME content to the content server	HTTP
PRS-14	Communication between the content server and the watcher to retrieve content stored and referenced by the Presence Source	HTTP
PRS-15	Storage and retrieval of content in/from the Presence server	HTTP

client' – a watcher – with the SIP/IP Core, in a way very similar to how POC-1 interface connects the PoC client with the SIP/IP Core. This consideration can be applied to interfaces PRS-2 to PRS-12.

PRS-13 interface is mainly used by Presence Sources to store content in a generic content server. Generally, HTTP PUT operations are performed over PRS-13. This is used to store user-related information such as icons, pictures, music or any other type of media or content that can be properly addressed via an HTTP URL.

PRS-14 interface is used by watchers to retrieve content stored in the Presence server. There are two main usages of PRS-14:

- Retrieval of Presence documents. In some cases, a Presence server sending a NOTIFY message to subscribed watchers may decide not to include a PIDF document in the payload section, but provide an HTTP URL to retrieve that document. This approach is used to save *signalling bandwidth* in the SIP network, and can be optionally implemented by some Presence servers. This content indirection mechanism is documented in [23].
- Delivery of media content pushed by a Presence Source. In this case, the PIDF-RPID document may contain links to content stored in the content server. Such links may point to media files or pictures, which can be retrieved by the watcher to get an enriched or more complete picture of the status of a given Presentity. Implementation of this feature with the `<rpid:status-icon>` element is shown in section 4.2.4.2.

4.3.3 Processing Presence Information

It would be simplistic to think that the Presence server behaves only as a proxy of Presence information between Presence Sources and watchers. In this section we briefly describe the set of tasks that the Presence server performs since a publication is received from a Presence Source until a notification is sent to watchers subscribed to the corresponding Presentity. The whole process is shown in Figure 4.9.

Presence Sources update Presence information by sending a SIP PUBLISH message that contains Presence information. Observe that, in general, there may be one or more Sources publishing Presence information on behalf of the same user. As we will see in section 4.6.2, a common configuration may consist on the end user device and the PoC Application Server behaving as Presence Sources for the same subscriber.

In general, Presence Sources publish Presence information about a Presentity whenever a relevant action is detected. This action may include user interaction with a device (e.g. activating the *Do Not Disturb* mode in a PoC handset) or may be triggered automatically (e.g. a Location Server detects that a user arrives at her home). Refer to Figure 4.3 for an example showing the two types of Presence Sources.

Given the potential multiplicity of Presence Sources capable of publishing Presence information about the same Presentity, the first function that the Presence Server must perform is *composition* of Presence information. The goal of composition is to correlate all received information and to produce a single Presence document that contains all pieces of information available at a given time. This is a complex function where the server determines which elements must be merged or kept separate. At this stage, the server must also solve any divergence, and decide – in case of conflicting information published by different Sources – which

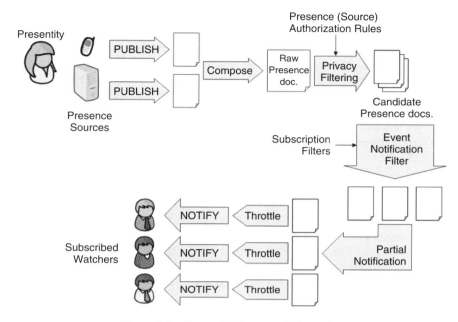

Figure 4.9   Processing Presence information

Presence information takes precedence (i.e. which are the most reliable Presence Sources). References [10, 11] provide a set or rules to perform composition. OMA [24] also defines a clear procedure to sort publication conflicts and ensure that the composition operation produces a consistent and coherent output.

The outcome of the composition function is a raw Presence document that consolidates all information received from several Presence Sources. All the information contained in the raw Presence document relates to one and only one Presentity.

Taking the raw Presence document as a basis, the Presence Server must generate a filtered version of it that is ready to be sent to each subscribed watcher. This is a key step, which ensures that each watcher only receives the set of information that the Presentity has allowed her to see. In order to perform this filtering function the Presence Server applies the *Presence Authorization Rules* which are applicable at each moment.

A Presence Authorization Rules document is an XML encoded document that basically defines who is authorized to receive what Presence information. This document is managed by the Presentity (i.e. the user who has Presence information associated to her) and is stored in the Presence XDMS. Presence Authorization Rules are discussed in detail in section 4.5.

The result of the Privacy Filtering function is a set of *candidate Presence XML documents.* Each candidate document is suitable for delivery to a subscribed watcher, based on the initial raw Presence document and the authorization rules which are applicable to that watcher.

Once the authorization filter has been applied, (we can think this as a source filter) the Presence server may apply another filter that is set up by the destination (i.e. the watcher). Effectively, the watcher may request the server to filter Presence information and send updates only when information relevant to the end user has changed. This is called *Event Filtering*.

As an example, a user may request to receive information when PoC availability of a Presentity changes, but he or she may not be interested in receiving updates when IM availability is modified (e.g. the watcher may have a PoC-enabled device that does not support IM capabilities). Event Filtering is activated by the watcher during subscription initialization: the SUBSCRIBE message sent by the watcher may contain an XML-encoded body that specifies a Presence information filter [25].

The outcome of the event filtering function is a Presence document that is ready to be sent to a particular watcher as described in [26]. In addition to event filtering, the Presence server may also use the partial notification mechanism. This procedure filters out information that is known to be already available at the watcher side. This mechanism is only applied if the subscribed watcher supports it.

Finally, the Presence server may apply a local policy that limits the rate at which notifications are sent to a watcher, in order to avoid flooding the network and the users with too many Presence-related messages. This feature is called *throttling*.

After all these stages have been completed (including potential extra delay added by the throttling function) the Presence Server sends a SIP NOTIFY message that contains a processed Presence document suited for each subscribed watcher of a given Presentity. This process is performed for all Presentities served by the Presence Server and for all publication messages received from all Presence Sources.

4.4 The Resource List Server

The Resource List Server (RLS) is an IETF-defined functionality [27], which OMA reuses and maps into the Presence architecture [22]. In fact, the RLS delivers one of the most important optimizations to support Presence services over wireless networks. The main goal of the RLS is to support a single subscription mechanism that lets watchers use one SIP SUBSCRIBE message to request Presence information about a list of Presentities, as opposed to having to perform individual subscriptions to each Presentity independently.

We can understand the basic functionality that the RLS offers by examining Figure 4.10.

Observe that Watcher A is interested in subscribing to Presence information about Presentities B, C and D. However, instead of sending three SIP SUBSCRIBE messages to the Presence Server, the Watcher performs the following procedure:

- First, Watcher A creates a Resource List document and stores it in the RLS XDMS (see section 4.5). This document stores a list of the URI's about which Watcher A is interested in receiving Presence information. The process of creating the list includes assigning it a SIP URI that points to the XML document (e.g. sip:list@wirelessfuture.com).
- After creating the Resource List document, Watcher A sends a SIP SUBSCRIBE message towards the RLS, and uses the SIP URI of the Resource List in the Request Line of the request (the Request Line will look like: `SUBSCRIBE   sip:list@ wirelessfuture.com SIP/2.0`).
- When processing this request, the RLS will retrieve the Resource List referenced by `sip:myList@wirelessfuture.com` from the RLS XDMS. The RLS will place an individual subscription against each URI in the list. Therefore, the RLS subscribes on behalf of Watcher A to receive Presence updates about Presentities B, C and D.

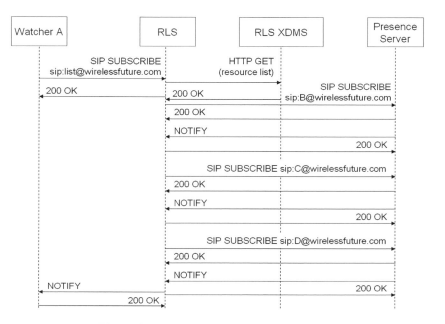

Figure 4.10   Usage of the Resource List Server

All subscriptions issued by the RLS are sent against the Presence Server. Observe that the Presence Server does not need to care whether a SIP SUBSCRIBE request comes from the end user of from the RLS: the RLS hides all the particularities of the Presence list functionality.

* Finally, after receiving all Presence information (e.g. PIDF and RPID documents) about each entry in the Resource List, the RLS sends a SIP NOTIFY message towards Watcher A. This message aggregates a multiparty body, that consists of
  ○ A *Resource List Meta Information* (RLMI) XML document, that sumarizes the status of all back-end subscriptions (between the RLS and the Presence Server) [27].
  ○ A set of Presence documents summarizing the available Presence information about all successfully subscribed Presentities.

The mechanism described above performs a significant reduction of the signalling overhead exchanged over the air interface, since the number of SIP messages is kept at four regardless of the size of the list (when no RLS is used the number of exchanged messages equals 4 × N, where N is the number of Presentities in the list). However, the size of the final SIP NOTIFY message grows significantly, as it will deliver one RLMI document and N Presence documents in a single SIP message.

4.5 XDM Presence Applications: Presence Policies and Resource Lists

There are two XDMS nodes defined by the OMA Presence architecture [22]: the Presence XDMS and the RLS XDMS. The XCAP application supported by the Presence XDMS is the

Presence Authorization Rules [28], while the RLS XDMS supports the *Presence List* XCAP application [29].

4.5.1 Presence Authorization Rules

Presence subscription authorization is an important topic: since Presence documents may convey sensible information, it is important to ensure that the end user (i.e. the Presentity) has full control over who is allowed to receive Presence information about him, and thus control how Presence subscriptions are handled.

With this idea in mind, the IETF has defined a general framework to let users (Presentities) decide how they want their Presence information to be disseminated towards watchers. This work is specified in the *Presence Authorization Rules* draft [30], which in turn reuses the general *Policy Management* framework defined by [31].

Taking IETF Presence Authorization Rules as a basis, OMA has defined the OMA *Presence Authorization Rules* XCAP application. OMA Presence Authorization Rules are basically an adaptation of IETF rules, but due to minor misalignments between OMA requirements and [30], a new XCAP AUID was defined (`org.openmobilealliance.pres-rules`), instead of reusing standard `pres-rules` as defined by the IETF Presence Authorization Rules.

In particular, OMA defines some extensions to [30] (e.g. usage of external URI lists stored in the Shared XDMS, definition of default rules or certain constraints into how rules should be interpreted), which lead OMA to define a new AUID.

The OMA Presence Authorization rules application [28] specifies that each Presentity may store a single XML document in the users' tree of the Presence XDMS (named 'pres-rules'). We can think a Presence Authorization rules document as a sequence of zero or more `<rule>` elements. Each `<rule>` element contains:

- Zero or more `<condition>` elements, which define under which conditions a rule should be applied (a condition may consist on matching an identity or a list of identities).
- Zero or more `<action>` elements, which detail which are the actions to be implemented when a condition is met (e.g. the action may consist on allowing or blocking a Presence subscription when the condition is met)
- Zero or more `<transformation>` elements, which define which transformations should be applied to a Presence document before delivering it to the watchers that fall under the condition element.

For each rule OMA makes the logical distinction between the `<condition>` – `<action>` section and the `<transformation>` section. Since each combination of *conditions* and *actions* defines whether a subscription is actually accepted or not, OMA defines this section of the Presence Authorization Rules document as the *Subscription Authorization Rules*. On the other hand, since the *transformation* section defines what actual Presence is shared once a subscription is accepted, this section of the document is defined as the *Presence Content Rules*.

The following picture outlines the scope of OMA Presence Authorization Rules.

In Figure 4.11 we can see how a user (e.g. John) has defined a set of Presence Rules for potential watchers subscribing to her Presence information, namely:

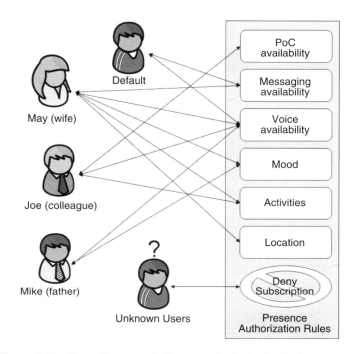

Figure 4.11   Example usage of a Presence Authorizarion Rules document

- Mary (John's wife) is able to receive John's Messaging availability, as well as his current activities, mood and location.
- Joe is a colleague that works with John. He is able to receive Presence information about his availability for the corporate PoC service they use to communicate regularly. Joe also receives information about John's activities.
- John's father receives information about regular Voice availability[5] and mood.
- John does not allow access to his Presence information from any unknown Watcher hiding her identity.
- In case none of the above conditions (identity matching) are met, the Presence service will deliver John's availability for Voice and Messaging sessions, as the default Presence Authorization Rule (observe that this rule applies to authenticated users that do not hide their identity).

Observe that when receiving a SIP PUBLISH message from a Presence Source, the Presence server must apply the available Subscription Authorization and Content Rules to reformat the Presence document into an authorized version for a given watcher. After performing this *source* filtering (managed by the Presence Source), it may apply the Event Notification Filtering to ensure that Presence information is relevant to the watcher (see Figure 4.9).

---

[5] In fact, version 1 of the OMA Presence enabler does not specify support for publication of availability for the regular (e.g. circuit switched) voice service. Its usage in this section is for example purposes only.

4.5.2 The Watcherinfo Event Package: Reactive Authorization

In addition to Presence Authorization Rules stored in the XDMS, the OMA/SIMPLE service supports another Presence authorization feature. Effectively, in the absence of a default rule or document in the Presence XDMS, a Presence Source may use a mechanism, so that whenever a new watcher intends to subscribe to Presence information about a user, he or she receives a notification message. The end user may then manually decide whether the subscription can be accepted or rejected.

This mechanism is based in the SUBSCRIBE – NOTIFY framework as well. In this case, Presence Sources can subscribe to the `presence.winfo` event package. Once the Presence Source subscribes to this package, the Presence server will send a NOTIFY message every time a new watcher wishes to subscribe (using the `presence` package) to receive Presence information about the Presentity.

This mechanism is called *reactive authorization*, as opposed to proactive authorization, which is performed when an applicable or default rule exists in the Presence Authorization Rules document. OMA Presence, however, does not fully specify how reactive authorization is to be implemented after receiving a notification of a new subscription request (i.e. which mechanism the Presentity should use to allow/block the subscription). As an example, a service provider could request sending an encoded text message to a certain address in order to define how to handle the subscription.

The presence.winfo package is defined in [34].

4.5.3 Presence Lists

We have explained in section 4.4 that the RLS supports subscription to Presence lists. As the reader may imagine, the RLS XDMS is used to store Presence list documents. The RLS retrieves lists stored in the RLS XDMS when it receives a SIP SUBSCRIBE message that includes the address of a Presence List in the Request URI. The Resource List is an IETF defined XCAP application [32], that is reused by OMA. Documents in the RLS XDMS are stored under the `rls-services` AUID tree.

As with any other XCAP application dealing with URI lists, it is possible that Presence lists contain links to external URI lists stored in the Shared XDMS. This is in fact the recommended application when the actual list of contacts is to be shared across several SIP/XCAP services.

In order to simplify client implementation, OMA defines that the XDMC[6] shall store all Presence lists under a single XML document, named `index`, under the `rls-services` tree. Within the index document, different Presence Lists can be easily distinguished based on the unique list name attribute assigned to each one.

The reader could think that there are many similarities between a Presence List document stored in the RLS XDMS and a URI List document stored in the shared XDMS, as described in chapter 3. This assumption is actually true, but the following differences apply:

---

[6] Remember that the XDMC is the logical entity that manages all XML documents stored in any XDMS, on behalf of the end user. Thus, any client device supporting Presence features must incorporate XDMC capabilities, to let the user manage Presence Authorization Rules and Presence Lists XML documents, stored in the Presence XDMS and the RLS XDMS respectively.

1. A Presence List must have a SIP URI associated to it, to be used in the SIP SUBSCRIBE request. A URI List has a name attribute (a token) associated to it, but it does not have any SIP URI attribute.
2. A Presence List may incorporate a `<package>` element that defines what SIP application makes usage of that list. The only *package* currently defined for Presence Lists stored in the RLS XDMS is `presence`.
3. A URI List is an application-agnostic XML document that may be reused by other application-specific XML documents (PoC groups, Presence Lists, Presence Authorization Rules). On the other side, the Presence List application is defined in scope of the Presence enabler, and cannot be reused by other services.

An example Presence List document is shown below. Observe the presence of the `<package>` element and the uri attribute of the `<service>` element.

```
<?xml version="1.0" encoding="UTF-8"?>
<rls-services xmlns="urn:ietf:params:xml:ns:rls-services"
              xmlns:rl="urn:ietf:params:xml:ns:
              resource-lists">

<service uri="sip:my_buddies@wirelessfuture.com">
   <list name="my_buddies">
      <rl:entry uri="sip:joe@wirelessfuture.com"/>
      <rl:entry uri="tel:5678;phone-context=+43012349999"/>
      <rl:entry uri="sip:mary@wirelessfuture.com"/>
      <rl:entry uri="sip:barbara@wirelessfuture.com"/>
   </list>
   <packages>
      <package>presence</package>
   </packages>
</service>
</rls-services>
```

4.6 Enhancing PoC User Experience with Presence Capabilities

4.6.1 Presence Enabled PoC Buddy List

The OMA Presence enabler can be used to enrich the PoC user experience. Effectively, PoC subscribers can use a Presence enhanced 'buddy list' to determine the best and least intrusive way to communicate with their individual contacts and groups. It is possible to deploy PoC without any Presence support. This option, however, may not be the most widely adopted. The following picture shows some examples where the Presence capability actually helps to build a richer enhanced user experience to the PoC service.

Let us assume that John has a PoC handset that incorporates a Presence-enabled buddy list. John may use Presence information to ensure that PoC communications with his personal and professional contacts are successful and non-intrusive. Consider the following three use cases:

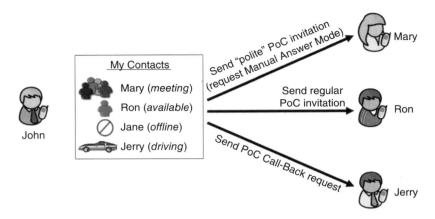

Figure 4.12 A Presence enhanced PoC buddy list helps John better tailor his PoC communications

1. John wants to talk to Mary. The PoC buddy list shows she is involved in a meeting. John would like to talk to her instantly, but in order not to disturb her, he sends a PoC session invitation requesting manual acceptance mode be used. Mary may either accept the call (if she is involved in an informal meeting) or reject it (if the meeting has higher importance than John's PoC call).
2. John wishes to talk to Ron, who happens to be 'available'. He sends a regular PoC session invitation message. The message will be accepted manually or automatically, based on the answer mode in use when the invitation arrives to Ron's handset.
3. John sees that Jerry is currently driving home. Thus, he will most likely not be able to accept any incoming PoC call. John sends a Call-Back Request message (PoC Instant Personal Alert) so that when Jerry arrives home can PoC him back to set up a meeting tomorrow morning.

Observe that depending on the actual set of rich Presence information displayed in John's handset, other interesting use cases are feasible.

Strictly speaking, the PoC service can be deployed without Presence capabilities. However, in such a case the user would interact with the service in a less controlled or *blind* way. As indicated above, Presence helps to use the most suitable service capability at each moment (e.g. PoC session, Call-Back request, Answer Modes), thus ensuring that interactions among users take place in a convenient way both for calling and called parties. We can conclude that Presence enhanced PoC service delivers a higher degree of user satisfaction when compared to a PoC service deployed without Presence features.

The ability of the Presence service to enhance the user experience will become even more aparent when version 2 of the OMA Presence enabler reaches maturity and commercial availability. Effectively, the Presence enabler will integrate a richer set of information, covering services such as SIMPLE, Text and Multimedia Messaging, Conferencing, Videocalling, PoCv2. The user will have at her hands a powerful tool seamlessly integrated with its Contact List application (possibly distributed and shared across several devices), that will let her have full control over all her multimedia communication services. Presence may then become a horizontal enabler and entry point for SIP and non-SIP-based applications in use in handsets, PDAs, smartphones or laptop devices.

In addition to the basic PoC – Presence relationship described above (Presence enabled buddy list), there are other more subtle ways in which the PoC and Presence enablers can interoperate. We describe one such case in the next section.

4.6.2 Interworking Between the PoC and the Presence Services

When designing the PoC and Presence enablers it became evident at some point that the set of information exchanged when the PoC client updates its PoC settings is redundant with the set of information that the same client device acting as a Presence Source exchanges with the Presence server. Basically, when a user changes her PoC settings (e.g. a user activates the Incoming Session Barring – ISB – flag) the following actions occur:

- A SIP PUBLISH message is sent to the PoC server (via the SIP/IP Core) publishing the new PoC setting (ISB activated). This is required, so that the PoC server blocks subsequent incoming session invitations to the user.
- In parallel, a SIP PUBLISH message is sent to the Presence server (via the SIP/IP Core) including a PIDF document that updates the Presence status (e.g. *unavailable for PoC sessions*). This is required so that the Presence server notifies subscribed watchers of the new status of the publishing user.

Observe that even though the two PUBLISH messages are not identical (the first one includes a PoC Settings XML document, and the second one a PIDF XML document), they contain basically the same information. In addition, both messages traverse the wireless (potentially narrowband, error prone, slow) link. As an optimization to this inefficiency, there is an optional feature, described by OMA, which lets the PoC server act as a Presence Source on behalf of the end user. When this feature is available in the PoC server, the second PUBLISH message is not sent by the client. Instead, the PoC server communicates any PoC settings change to the Presence server. Observe that with this configuration the service is more efficient, because only one PUBLISH message is exchanged over the wireless link. This feature is depicted in Figure 4.13.

1. Upon device activation, the PoC client registers a PoC identity in the SIP/IP Core. The PoC server may receive a notification from the SIP/IP Core informing that the PoC user is now registered.
2. After successful registration, the PoC client publishes the PoC service settings (e.g. available for PoC sessions and IPA). Not all SIP/IP Core networks may notify application servers about the registration status of their users. Hence, this step is required so that the PoC server is aware of user availability, and to ensure that the proper session initiation mechanisms – manual vs. automatic – are used.
3. After completion of step 2, the PoC server knows:
   a. The registration state of the user
   b. Her PoC availability settings (availability for PoC sessions, availability for IPA)
       At this stage, the client device could send a new PUBLISH message to provide this same information set to the Presence server. Instead, it is the PoC server acting as a

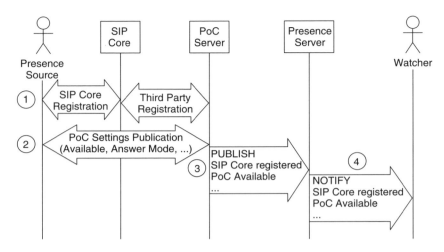

Figure 4.13   PoC Server publishing Presence information on behalf of the user

> *Presence Source on behalf of the user* that sends a PUBLISH message that includes all the information set that has been made available to the PoC server in steps 1 and 2.
4. The Presence server may now send a SIP NOTIFY message to all subscribed watchers updating the status of the Presence Source that motivated the Presence publication sent in step 3.

Observe that, for sake of clarity, in Figure 4.13 we have not depicted an actual message flow – not all message flows traverse the SIP/IP Core in the figure.

Activation of this feature (PoC server publishing Presence information on behalf of the user) is specified in annex C of the OMA PoC Control Plane specification [33]. When this feature is enabled, it is important to ensure that the client device is aware of it, so that Presence PUBLISH messages are not sent (otherwise, this feature is useless). Client awareness is ensured by activating a client configuration parameter via device management operations, as specified in annex B of the same specification [33].

In addition to the ability of behaving as a Presence Source, [33] specifies that the PoC server may optionally behave as a watcher as well (i.e. subscribe to Presence information about served PoC users). However, it is not clear at the time of writing whether this functionality may be useful at all or not (given that the PoC client must send a SIP PUBLISH message to the PoC server anyway). Thus, we do not expect that this feature becomes widely available at first stages.

4.7 Summary, Conclusions and Some Final Comments about the Presence Service

We have seen in this chapter that OMA Presence develops a mobile-friendly Presence service, which enriches user experience and complements other service offerings such as PoC, IM, VoIP or multimedia group communications in general. There are, however, several additional items to take into account when considering real deployment of the service. We briefly discuss some of these topics without the aim of being exhaustive but, rather, to point out some areas for further discussion when thinking about wide scale deployment of a mobile Presence service.

The first topic is to understand which is the best strategy to launch Presence services. It seems clear that basic Presence enriches overall user experience of mobile subscribers. On the other hand, basic Presence does not provide a communication service per se, as other services (e.g. PoC, IM, Voice, Streaming) may do. From that point of view, Presence *enhances* user experience associated to services, but *does not provide a service* offering in itself.

Consequently, cellular operators may consider launching basic Presence services for free, together with other SIP-based multimedia services such as PoC or IM. Presence can be considered as a *horizontal enabler* that supports other services, which become Presence-aware. When thinking about this scenario there are still some questions to be answered. Consider the following two approaches that could be followed:

- A Presence enabled device implements a contact list application that displays Presence information about all entries in the list. From that *horizontal* contact list application the user may invoke several services such as PoC, IM and VoIP.
- Each application (e.g. PoC, IM) in the device implements a separate Presence enabled contact list. When the user enters each application, Presence information about availability for that particular application is displayed.

Observe that both approaches seem to be fully feasible in scope of OMA Presence. However, each option may lead to totally different design issues and user experiences. For example, both alternatives may differ in the way shared groups are defined and stored, and in the way event-filtering mechanisms are used. Finally, both implementations will generate different traffic patterns as well.

It is also worth observing that there are a number of areas within OMA Presence that still require further fine-tuning and actual experience gained from real deployments. In particular, those elements in the Presence service specified as *optional* will require additional efforts to ensure seamless interoperability across networks and devices. To name a couple of topics in this area we would like to point out the following:

- Configuring the whole set of optional tools to optimize Presence traffic requires that several elements in the end-to-end path work in a coordinated way (e.g. RLS, Event Filtering, Partial Publication and Notification, configuration of SIP SUBSCRIBE timers, SigComp).
- Several Presence data formats (PIDF, RPID, the Presence Data Model, the location extensions) may be used to deliver a rich set of Presence information. Basic interoperability should be ensured even in complex scenarios when several Presence format extensions are used.

The need to ensure consistency and avoid conflicting Presence information issues becomes even more critical when building complex Presence enabled applications, such as location based services, Presence triggered notifications, and device capability publications.

Finally, it is important to understand that GUI design is a key enabler of Presence service success, adoption and ease of use. Readers may be familiar with existing Presence applications available in the Internet domain. Other examples are shown through this text (e.g. Figure 4.1 or Figure 4.12). It is a challenge to combine:

a) Limited display capabilities, as those typically available in mobile devices
b) The ability to render complex information including several applications, mood, activities, location, willingness, . . .
c) A GUI design that aims for usability and clarity

Being able to find the right combination of these three constraints or design issues will pave the way for the success of a mobile Presence service enhancing multimedia group communications.

4.8 References

[1] OMA Instant Messaging and Presence Service (OMA IMPS v1.2.1): 'OMA IMPS Enabler Release Definition'; August 2005.
[2] OMA Work Item Document 'Group Management'; WID-0073, November 2003.
[3] OMA Presence and Availability Group: 'Liason Statement proposing work split 3GPP/3GPP2/OMA on Presence', OMA-PAG-2004-0328R02, August 2004.
[4] 3GPP Services & Systems Aspects Working Group 2: 'LS on proposed work split 3GPP/3GPP2/OMA on Presence', NP-040474, October 2004.
[5] M. Day, et al.: 'A Model for Presence and Instant Messaging', RFC 2778, February 2000.
[6] J. Peterson: 'Common Profile for Presence (CPP)', RFC 3859, August 2004.
[7] J. Rosenberg, H. Schulzrinne, G. Camarillo, A. Johnston, J. Peterson, R. Sparks, M. Handley and E. Schooler: 'SIP: Session Initiation Protocol'. RFC 3261, June 2002.
[8] A. Roach: 'Session Initiation Protocol (SIP)-Specific Event Notification', RFC 3265, June 2002.
[9] J. Rosenberg: 'A Presence Event Package for the Session Initiation Protocol (SIP)', RFC 3856, August 2004.
[10] A. Niemi: 'Session Initiation Protocol (SIP) Extension for Event State Publication', RFC 3903, October 2004.
[11] H. Sugano, S. Fujimoto, G. Klyne, et al.: 'Presence Information Data Format (PIDF)', RFC 3863, August 2004.
[12] M. Day, et al.; 'A Model for Presence and Instant Messaging', RFC 2778, February 2000.
[13] M. Day, et al.; 'Instant Messaging/Presence Protocol Requirements', RFC 2778, February 2000.
[14] J. Rosenberg, 'A Data Model for Presence', RFC 4479, July 2006.
[15] H. Schulzrinne, et al.; 'RPID: Rich Presence Extensions to the Presence Information Data Format (PIDF)', RFC 4480, July 2006.
[16] J. Peterson, 'A Presence-based GEOPRIV Location Object Format', RFC 4119, December 2005.
[17] OpenGIS: 'Open Geography Markup Language (GML) Implementation Specification', OGC 02-023r4, January 2003.
[18] H. Schulzrinne, H. Tschofenig, J. Morris, et al.: 'Geolocation Policy: A Document Format for Expressing Privacy Preferences for Location Information', Internet Draft (work in progress), February 2007.
[19] H. Schulzrinne, 'CIPID: Contact Information in Presence Information Data Format', RFC 4482, July 2006.
[20] H. Schulzrinne: 'Timed Presence Extensions to the Presence Information Data Format (PIDF) to Indicate Status Information for Past and Future Time Intervals', RFC 4481, July 2006.
[21] M. Lonnfors, K. Kiss, 'Session Initiation Protocol (SIP) User Agent Capability Extension to Presence Information Data Format (PIDF)', Internet Draft (work in progress), IETF, July 2006.
[22] OMA Presence (Presence v1.0.1): 'Presence SIMPLE Architecture'; November 2006.
[23] E. Burger: 'A Mechanism for Content Indirection in Session Initiation Protocol (SIP) Messages', RFC 4483, May 2006.

[24] OMA Presence (Presence v1.0.1): 'Presence SIMPLE Technical Specification'; November 2006.

[25] H. Khartabil, E. Leppanen, M. Lonnfors, J. Costa-Requena: 'Functional Description of Event Notification Filtering', RFC 4660, September 2006.

[26] H. Khartabil, E. Leppanen, M. Lonffors, J. Costa-Requena: 'An Extensible Markup Language (XML) Based Format for Event Notification Filtering', RFC 4661, September 2006.

[27] A. B. Roach, B. Campbell, J. Rosenberg: 'A Session Initiation Protocol (SIP) Event Notification Extension for Resource Lists', RFC 4662, August 2006.

[28] OMA Presence (Presence v1.0.1): 'Presence XDM Specification', November 2006.

[29] OMA Presence (Presence v1.0.1): 'Resource List Server (RLS) XDM Specification', November 2006.

[30] J. Rosenberg: 'Presence Authorization Rules', Internet Draft (work in progress), IETF, October 2006.

[31] H. Schulzrinne, H. Tschofenig, J. Morris: 'Common Policy: An XML Document Format for Expressing Privacy Preferences', RFC 4745, February 2007.

[32] J. Rosenberg, 'Extensible Markup Language (XML) Formats for Representing Resource Lists', Internet Draft (work in progress), IETF, February 2005.

[33] OMA Push-to-Talk over Cellular (PoC v1.0.1): 'PoC Control Plane'; November 2006.

[34] J. Rosenberg: 'A watcher Information Event Template-Package for the Session Initiation Protocol (SIP)', RFC 3857, August 2004.

[35] OMA Presence (Presence v1.1): 'Presence SIMPLE Architecture'; September 2007 (Work in Progress).

5

Deploying Group Communication with IMS

5.1 Introduction

The application of multimedia group communication to the mobile domain involves two approaches. The first approach is to explain the integration of OMA PoC, XDMS and Presence with SIMPLE on top of SIP/IP Core and with related IP based service entities. The second approach is to give examples of group communication through a series of use case signalling charts and entity handshakes.

This chapter introduces the reader to group communication deployment on top of an SIP/IP Core. It specifically addresses the 3GPP IP Multimedia Subsystem (IMS) and refers to practical issues such as PoC charging, device management and optimizing the radio network. Section 5.2 outlines IMS concepts and corresponding PoC and XDM reference points. Section 5.3 explains the relation between IMS User Identity Management and group communication. Section 5.4 gives examples on connectivity between PoC and Presence with IMS, and between PoC domains. Section 5.5 gives a detailed description on PoC charging and covers online (prepaid) and offline (post paid) models. Section 5.6 lists the parameters required to provision the UE with the network services. Section, 5.7 specifics the network parameters designed to improve PoC subscribers' QoS.

5.2 3G IP Multimedia Subsystem (IMS) Concepts

The 3GPP *IP Multimedia [Core Network] Subsystem* (the IMS) defines a complete architecture for the delivery of IP-based multimedia services to the end user. 3GPP specifies IMS as a set of logical functions interconnected via well-defined standard interfaces. As with OMA architecture, several logical functions can be combined in a single physical entity, and vice-versa. SIP is the main signalling protocol that supports user-to-user and user-to-network services over IMS. Given the relevance of the IMS paradigm for the evolution of the 3G network, and the alignment between 3GPP and OMA activities, it seems logical to assume that the standard PoC architecture adopted by cellular operators will consist on a PoC Application

---

Multimedia Group Communication. Andrew Rebeiro-Hargrave and David Viamonte Solé
© 2008 John Wiley & Sons, Ltd.

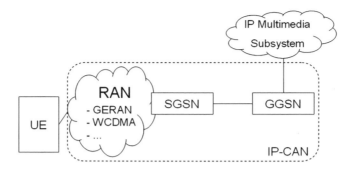

Figure 5.1   IMS implementation over the cellular IP-CAN

Server deployed over a 3GPP compliant IMS infrastructure that implements the 'SIP/IP Core' functionality.

Besides basic VoIP capabilities defined by 3GPP Release-6, PoC and Presence are the first SIP-based standardized services that should be seamlessly supported by IMS. Therefore, this and subsequent sections aim at providing the reader an overview of how OMA PoC services can be successfully deployed over a 3G IMS infrastructure. A certain degree of understanding of the IMS concepts and ideas is assumed. Readers that are not familiar with the IMS paradigm are referred to [1][2] as two very good sources of information to get an overview of the IMS technology.

The IMS provides services to mobile users over the existing 2.5G or 3G infrastructure. Since the IMS architecture is basically access agnostic, the only assumption the IMS makes is that the underlying network is IP-based, so it makes the abstraction that such network is an *IP Connectivity Access Network* (IP-CAN). Figure 5.1 shows how such IP-CAN is implemented in a generic 3GPP cellular network.

In Figure 5.1 we can see that two main building blocks build up the IP-CAN when implemented over a 3GPP cellular network [4]:

- The *Radio Access Network* (RAN) supports cellular coverage through the usage of base stations deployed over a certain geographic area. We may talk about GPRS/EDGE RAN (GERAN) or UMTS RAN (UTRAN), depending on whether 2.5G or 3G technology is used to support the wireless link.
- The packet core network is generally built up by the connection of two logical nodes: the SGSN, which supports mobility management, and the GGSN, which supports connectivity to external IP networks (such as the IMS domain). Both nodes support other features such as charging or QoS management. Each operator may have several SGSNs and GGSNs depending on the actual traffic and subscriber volumes. The GGSN offers the standard *Gi/Mb*[1] reference point towards external networks. This is the entry point to the IMS infrastructure.

---

[1] The *Gi* interface is defined in [4]. It is the IP-based interface which connects the core network providing services to cellular users, to an external IP-network. For GPRS-based Core-Networks, the Gi interface is supported by the GPRS Gateway Support Node (GGSN). SIP traffic exchanged between a UE and the IMS is seen as plain 'User Plane' traffic, from the cellular network point of view. Thus, the GGSN sees the IMS in the same way as any other IP-based network (althoug specific QoS treatment may be assigned to SIP signalling trafffic, to ensure reliability and fast delivery). The *Mb* interface provides functionality similar to Gi, but Mb is IPv6 specific (Gi supports IPv4) only.

IMS defines a set of logical functions that proxy SIP signalling between two or more SIP endpoints. These nodes are the *Call Session Control Function* (CSCF) elements. There are three main CSCF functions defined by 3GPP:

- The *P(roxy)-CSCF* is the first entry point between the mobile SIP client and the IMS core network. It supports signalling compression over the air interface (in order to reduce the transfer time for signalling messages) and establishes a security association to encrypt SIP signalling exchanged with the client. The interface between the P-CSCF and the UE is called *Gm*.
- The *S(erving)-CSCF* provides services to a particular IMS user, and behaves as a SIP registrar. The S-CSCF determines which and how services are provided to each user. To do so, the S-CSCF retrieves the user profile from the *Home Subscriber Server* (HSS). The S-CSCF may provide basic peer-to-peer services to a user. When a more complex service is required, SIP requests are terminated at an Application Server (such as the PoC Server). The S-CSCF is the single entry point between Application Servers and the IMS. All SIP messages related to a user traverse the S-CSCF (by making use of the SIP Record-Route header), so that this is the central point for service control.
- The *I(nterrogating)-CSCF* is capable of performing routing decisions based on queries to the HSS. The I-CSCF is responsible for routing initial registration messages to the right S-CSCF (it queries the HSS to determine which S-CSCF should serve the user), and it also manages traffic that is to be routed to external IMS domains, or traffic related to roaming users.

The IMS defines a central database that stores user related data. The *Home Subscriber Server* (HSS) is a key element within the IMS architecture. The S-CSCF and the I-CSCF query the HSS to retrieve user and service information: the S-CSCF downloads user profile information to authenticate the user and to determine how to serve her. Among other operations, the I-CSCF queries the HSS to determine how to route a message towards an external SIP network.

The combination of CSCF nodes and the HSS represent the core of the IMS. However, additional functions are required to support a rich and interoperable set of services to end users. The following paragraphs briefly describe the additional elements depicted in Figure 5.2

- The *Media Gateway Control Function* (MGCF) and the *IMS-Media Gateway* (IM-MGW) provide the necessary means to interoperate with external circuit-switched networks. These nodes support establishment of voice/video calls between IMS users and regular PSTN/ PLMN users.
- The *Media Resource Function Controller / Processor* (MRFC and MRFP) support media resource functions such as media mixing and media replication in conferencing applications. They may support additional services such as media transcoding or generation of multimedia announcements.
- Routing of SIP messages in the IMS domain is based on the SIP URI. When an IMS user calls another user in a circuit switched network, a TEL URL[2] is generally used instead [7], and there may not be any SIP URI associated to the called user. In such case, the IMS does

---

[2]The TEL URL format [7] is used to refer to legacy telephone devices in the SIP domain. As an example, when an IMS user wishes to contact a legacy GSM subscriber, the TEL URL will contain the GSM number (i.e. the MSISDN) associated to that mobile user. In its more general form, the TEL URL contains the full E.164 encoding of a telephone number, including the associated *country code* and prepended by a '+' sign. Abbreviated forms can be used, provided that a mechanism to uniquely determine the proper telephone number can be inferred. IMS subscribers may be assigned a TEL URL as well in order to enable full interoperability with legacy non-IMS subscribers.

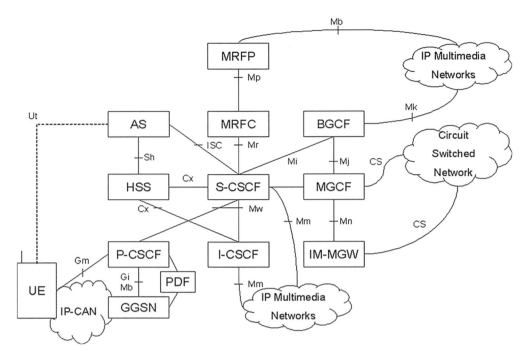

Figure 5.2   3GPP IMS architecture

not know how to route the message further. In order to cope with routing decisions based on TEL URL identifiers, the *Break-out Gateway Control Function* (BGCF) is used: the S-CSCF receiving such request will route it towards a BGCF node. The BGCF may decide to route the call towards a BGCF in an external IMS network, or to redirect it to a MGCF node, to terminate the call into an external CS network.

- The *Application Server* (AS) is a SIP entity that provides services to the end user. Application Servers are connected to the S-CSCF via the standard *IMS Services Control* (ISC) interface. When a SIP message requires interaction with an AS, the S-CSCF routes that message towards the right Application, which hosts one or more SIP services. In general, there are three ways in which an AS may be configured to provide SIP services, namely:
  - As a *SIP Proxy*. When an AS receives a SIP message, the AS may change some header values and insert new information. The resulting message will be routed further. The SIP request will be terminated at another network entity (e.g. a SIP user agent). An AS behaving as a SIP Proxy does not proactively process a request: it only proxies SIP requests and responses, possibly logging some data related to the session. A generic SIP messaging server could be an example SIP Proxy service.
  - As a *User Agent*. Application Servers configured this way terminate SIP dialogs. Hence, when a SIP request arrives to an AS, it provides a SIP answer message. A voicemail application can be implemented using an AS configured as a User Agent.
  - As a *Back-to-Back User Agent* (B2BUA). B2BUA are user agents that manage two different call legs. The AS terminates one leg for the originating party and initiates a second call leg against one (or more) terminating party(ies). A conferencing server that dials-out

towards a set of participants, upon conference initiation, could be an example B2BUA. When managing PoC sessions, the PoC Server generally connects to the IMS as a B2BUA.

- The *Policy Decision Function* (PDF) / PCRF supports policy management functions between the control layer (i.e. the IMS) and the user plane (i.e. the traffic flowing through the GGSN). By connecting these two layers, the PDF lets the Operator control network resources and policies and apply gating functions.

We have seen that the *Gm* interface is the entry point to the IMS for SIP-based clients. On the other hand, the *ISC* is the main communication interface between SIP applications (Application Servers) and the IMS. We can say that these are two of the most important 'external' IMS interfaces: they are exposed to external applications residing in the UE or in the AS, respectively. In addition to the *Gm* and *ISC* reference points, 3GPP defines the *Ut* interface as a direct logical connection between an IMS UE and an AS. The Ut interface is based on HTTPS transport, and is used by end users and client applications to manage settings of IMS applications. As an example, XDM-3 reference point defined by OMA to manage XCAP documents is mapped into the 3GPP Ut interface, when the OMA XDM service is deployed over a 3GPP compliant IMS core system.

5.3 OMA PoC over IMS

5.3.1 Mapping of OMA and 3GPP IMS Reference Points

In chapter 2, we presented the Push-to-Talk architecture as a service which is deployed over a SIP/IP Core that provides capabilities such as user authentication, service discovery and routing of SIP signalling. By inspecting Figure 5.2 and the PoC architecture, we can easily map OMA PoC reference points into the corresponding IMS interfaces:

- OMA POC-2 (between the SIP/IP Core and the PoC server) reference point is implemented over the 3GPP ISC interface. For the PoC AS, the anchor point with the IMS is the S-CSCF. This interface supports third party registration, setup of originating and terminating SIP (i.e. PoC) sessions and delivery of SIP MESSAGE based PoC operations (i.e. Instant Personal Alert and Group Advertisement transactions).
- POC-6 (between the SIP/IP Core and the PoC XDMS) reference point is implemented over the 3GPP ISC interface. This interface supports subscription to changes in XML documents. Observe that the PoC XDMS is seen by the IMS as an application server, when it comes to supporting POC-6 functionality.
- POC-1 reference point is implemented over 3GPP *Gm* interface, which connects the IMS client in the UE with the P-CSCF. Observe that mapping OMA POC-1 to 3GPP Gm interface means that an OMA PoC and IMS compliant handset must simultaneously support features such as establishing a security association with the P-CSCF, implement *signalling compression* (SigComp) capabilities and IMS authentication based on *Authentication and Key Agreement* (AKA).
- The IP-1 reference point, defined by OMA for session signalling between networks, is implemented by 3GPP Mm reference points (defined between CSCF nodes that implement the interconnection function and pertain to different domains). Observe that the IP-1

reference point is not PoC-specific, and is defined as well for other enablers such as OMA Presence or OMA Instant Messaging.

• The Ut interface, defined for management of application data, is mapped into OMA XDM-3 reference point, which is used to enable communication between the XDMC and the XDM service (via the Aggregation Proxy), to let the user manage groups.

• The optional XDM-1 reference point maps into the Gm interface, to support subscription and notification of changes in XML documents.

Figure 5.3 shows a graphical display of the mapping between OMA PoC and 3GPP IMS reference points.

This mapping of OMA PoC reference points to 3GPP IMS interfaces is also applicable to the Presence and XDM architecture. We can derive the correspondences of OMA Presence and OMA XDM reference points vs. IMS interfaces as well (these are not shown in the figure above for sake of clarity):

• PRS-1 and PRS-2 reference points are implemented over the Gm interface. This interface supports Presence watcher and Presence Source operations such as subscription/notification and publication of Presence information, respectively.

• PRS-3, PRS-4, PRS-6 and PRS-11 reference points are implemented over the ISC interface. These support communication between the IMS and the Presence server, the RLS, the Presence XDMS and the RLS XDMS respectively.

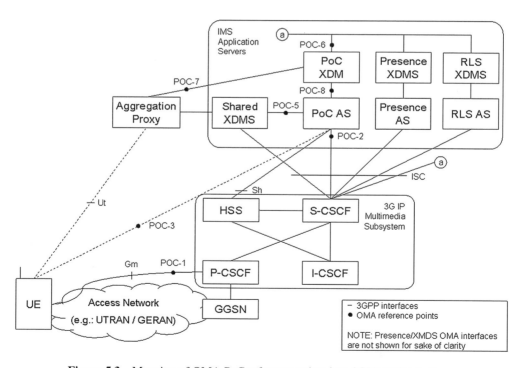

Figure 5.3   Mapping of OMA PoC reference points into 3GPP IMS interfaces

- Reference point XDM-2 is implemented over the 3GPP ISC (reference points XDM-1 and XDM-3 map into 3GPP Gm and Ut respectively, as explained above).

5.3.2 The PoC Server as a SIP Proxy

The PoC server may behave as a plain SIP Proxy in the following cases:

- When processing transactions based on the SIP MESSAGE method. This applies to the delivery of PoC Group Advertisement and Instant Personal Alert messages.
- When the Participating PoC function does not need to stay in the media path.

When the PoC server behaves as a SIP Proxy it basically checks validity of incoming messages and performs a routing decision to forward an incoming message towards its intended destination (e.g. a remote PoC user or a remote PoC server). The PoC server may reject an incoming request (e.g. if an invoked operation violates a SIP or OMA requirement). However, once a request has been proxied towards its destination, the PoC server does not have any additional control over how the transaction proceeds.

In the particular case of PoC session initiation, once the PoC server (behaving as a SIP Proxy) has forwarded a SIP INVITE message towards its next hop, it does not have any mechanism to influence how the PoC session evolves in the future. For example, the PoC server cannot initiate any mechanism to terminate a session leg (e.g. remove a participating user from a session).

5.3.3 The PoC Server as a B2BUA

A B2BUA is simply a concatenation of a SIP *User Agent Server* (UAS) and a SIP *User Agent Client* (UAC). The UAS terminates incoming requests, and the UAC initiates outgoing requests as a response to incoming messages processed by the UAS. SIP [8] does not define the type of relationship between incoming and outgoing SIP requests as managed by the B2BUA. For each particular application the behaviour of a B2BUA may be specified. In fact, OMA PoC defines how the PoC AS behaves as a B2BUA.

For example, when a user initiates an ad-hoc PoC call towards another user, there are actually two SIP sessions (actually, two SIP dialogs) being handled by the PoC server as a B2BUA:

- The *originating dialog*, between the calling PoC user and the PoC server that serves that user. In a certain sense, the PoC server is seen by the IMS as a regular SIP user agent, identified by the *PoC Conference Factory URI*.
- The *terminating dialog*, between the *originating* PoC server and the remote end (e.g. the called PoC user or another PoC server).

In order to handle sessions, the PoC server maps incoming RTP / TBCP packets and SIP transactions between both parallel SIP/PoC sessions. This is, effectively, the key behaviour of the PoC Server as a B2BUA.

The Controlling PoC function always behaves as a B2BUA, as it needs to handle all signalling and media related to a given PoC session. When the Participating PoC function stays in the media path, it behaves as a B2BUA as well.

Observe that if an operator wishes to have a tight control over how its subscribers use the PoC service, it will generally configure its PoC servers to always behave as a B2BUA when implementing the Participating PoC function. This is the only configuration that lets the operator support in-call session policies.[3] As an example, the PoC server may remove a user from a session when she has run out of credit to use the service.

5.4 IMS User Identity Management

PoC users are identified by their corresponding PoC address. It is possible that a PoC subscriber is assigned one or more PoC addresses (for example, a subscriber using her professional and personal profiles from a single device). In any case, the OMA PoC standard mandates that at least one of the addresses assigned to a PoC user be in the form of a SIP URI (*e.g. sip:joe.doe@wirelessfuture.com*). In addition to the SIP URI format, the TEL URL is the only additional format acceptable for PoC addresses. The Tel URL schema was defined to support identification of resources such as telephone numbers [7]. An example TEL URL address could be *tel:+35812323232*. SIP URI format is described in [8], and is the standard address schema used to route SIP messages within a SIP domain (e.g. the IMS and associated services). TEL URL is a format standardized in [8], and is used to support communication with legacy CS networks (which use the E.164 numbering plan, as the basis to identify users and route phone calls).

As the reader would expect, when the PoC service is run on top of the 3GPP IMS, each PoC address must be a valid IMS address as well, in the same manner as PoC subscribers are also IMS subscribers.

The IMS defines two types of user identities: the *IMS Private User Identity* (IMPI) and the *IMS Public User Identity* (IMPU). The PoC address must correspond to one of the IMS allocated Public User Identities. The IMPI is a private identifier only used internally between the home IMS network and the IMS subscriber, and accomplishes a role similar to that of the IMSI allocated in a legacy GSM *Subscriber Identity Module* (SIM). The IMPI is only used during the registration stage, and cannot be used for any further interaction with the IMS or any IMS service. IMS addressing paradigms are specified in [10].

In general, cellular subscribers equipped with IMS enabled handsets will access IMS services using their associated *IMS Subscriber Identity Module* (ISIM). Among other details, the ISIM stores the assigned IMPI and IMPU identifiers, and security keys used to authenticate the user and encrypt IMS signalling. The ISIM expands the SIM and USIM concepts associated to GSM and UMTS networks, to the IMS domain.

As an example, a PoC/IMS subscriber may have assigned the following respective IMPI and IMPU:

- Private User Identity: <234150999999999@wirelessfuture>
- Public User Identity: <sip:joe.doe@wirelessfuture.com>

---

[3] Although these policies are not clearly defined by OMA PoC [5.9], the PoC server may transparently implement them without causing any interoperability problem.

Both identities are stored in the ISIM card.

Consider now that the ISIM card storing the above identities is placed in a PoC enabled handset. The PoC client can retrieve the IMPU value from the ISIM and use it as the PoC address associated with the subscriber.

If we now consider the XDM service, we may recall that part of its functionality relies on non-SIP interfaces (e.g. 3GPP Ut or OMA XDM-3). The Aggregation Proxy is in charge of properly authenticating the XCAP user over such interface (hence, the need to carry user identity in XCAP messages). User identity is populated in the HTTP `X-XCAP-Intended-Identity` or `X-3GPP-Intended-Identity` headers of each XCAP request the XDMC initiates against the Aggregation Proxy (refer to chapter 3). Such *XCAP User Identity* (XUI) is retrieved by the XDMC from the IMPU stored in the IMSI as well. Hence, in general, the XUI is equal to the PoC address, and both are equal to one IMPU.

In case several PoC/XUI addresses have been allocated to a single subscriber, all addresses must be stored in the ISIM as valid IMPU addresses as well.

Observe that, by using the ISIM module and a properly configured PoC/XDMC client, the user is able to seamlessly access all her associated OMA services in a consistent way: the IMPU stored in the ISIM is used to feed the PoC and XUI addresses, so that all user related information is kept coherent across platforms and services (e.g. HSS, PoC subscriber database, XCAP users tree). Apart from the addressing information provided by the ISIM the PoC and XDM clients must be provisioned with configuration data as defined in [9] and [11] respectively. OMA device management capabilities can be used to provision such configurations. IMS provisioning parameters are self-contained in the ISIM.

The relationships between the IMS, PoC and XCAP addresses when accessing the service from an ISIM equipped device are shown in Figure 5.4.

Observe that several IMPU's can be associated to a single IMPI (hence, to a single subscriber). As an example, a user may have both a SIP URI and a tel URI addresses assigned as her associated IMPU's. Both addresses can be used to access the PoC service.

5.4.1 Access to PoC IMS Services Using a SIM/USIM Module

We have seen how ISIM-based access to the PoC service is quite straightforward. However, we can expect that for a relevant period of time, potential IMS subscribers will

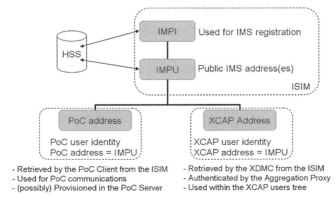

Figure 5.4   Relationship between IMS, PoC and XCAP addresses

not be upgrading their legacy SIMs or USIMs to an ISIM card. It is foreseeable that when a user upgrades her cellular handset to an IMS/OMA PoC device, she may expect to keep using the same SIM/USIM physical card she used in the past (e.g. in order to keep the contact list stored in the legacy card). In such case, 3GPP has defined a set of mechanisms that should let an IMS-capable device use information contained in legacy SIM/USIM cards, to successfully register in an IMS network and invoke IMS services regularly.

Since a SIM/USIM card does not store IMS addressing information, a mechanism has been defined in [10] to let an IMS client derive temporary IMS addressing information from identity information stored in a SIM/ISIM module. One of the important identity information stored in the SIM card is the *International Mobile Subscriber Identity* (IMSI). The IMSI is a private identifier stored in the SIM, and known to the GSM/UMTS network. The IMSI is not used to contact a cellular subscriber. Rather, the MSISDN (Mobile Subscriber ISDN Number) is used instead.

The IMSI is constructed by concatenating the following information set:

- The *Mobile Country Code* digits (e.g. 34)
- The *Mobile Network Code* digits assigned to the Service Provider (e.g. 609)
- The *Mobile Subscriber Identification Number* (MSIN) (e.g. 365436432).

The mechanism defined in [10] lets IMS enabled devices configured with a legacy SIM/USIM create a set of temporary IMS identifiers, so that the user is able to access IMS based services (such as PoC) even in the absence of the ISIM module. The procedure followed by the IMS client to create such temporary identifiers is described here:

1. The client creates a temporary IMS Private User Identity (IMPI) by concatenating:
   - The IMSI as the username part of the IMPU (e.g. 34609365436432).
   - The Home Network Domain name using a convention defined by 3GPP (e.g. ims.mnc34.mcc609.3gppnetwork.org).
2. The client creates a temporary IMS Public User Identity (IMPU) by prepending the 'sip:' string to the IMPI.

As an example, an IMS client may derive the following IMPI and IMPU from the IMSI stored in the SIM card:

- Private User Identity: <34609365436432@ims.mnc34.mcc609.3gppnetwork.org>
- Public User Identity: <sip:34609365436432@ims.mnc34.mcc609.3gppnetwork.org>

Once the IMPI/IMPU has been derived, the IMS client is ready to authenticate and register to the IMS. Observe that encoding the public identity based on the user's unknown IMSI leads to rather large and complex addressing schema. Subscribers are used to managing MSISDN digits, but not IMSIs. For this and for security reasons, the addresses obtained with this mechanism can only be used during the registration stage, and must not be used in any other SIP/IMS transaction. In particular, the user should not share the derived IMPU value with her colleagues as her contact IMS address, as the IMS will reject all incoming SIP sessions targeting such destination.

If the IMPU derived from the IMS is only used for registration and authentication purposes, subscribers accessing IMS/PoC services from a SIM equipped handset need to obtain 'usable' IMS identifiers by some means. In order to achieve this goal, the IMS inserts a special SIP header in the final registration response sent to the IMS client once the registration process is successful. This header contains a list of all implicitly or explicitly registered addresses that are provisioned in the IMS for the registering user. Consider for example the following simplified messages:

1. The PoC/IMS handset equipped with a SIM card sends the following REGISTER message:

```
REGISTER sip:ims.mnc34.mcc609.3gppnetwork.org SIP/2.0
From: <sip:34609365436432@ims.mnc34.mcc609.3gppnetwork.org>
To: <sip:34609365436432@ims.mnc34.mcc609.3gppnetwork.org>
```

2. After challenging the user and implementing the IMS AKA authentication, the S-CSCF sends a final successful response to the registration request:

```
SIP/2.0 200 OK
From: <sip:34609365436432@ims.mnc34.mcc609.3gppnetwork.org>
To: <sip:34609365436432@ims.mnc34.mcc609.3gppnetwork.org>
P-Associated-URI: <sip:joe.doe@wirelessfuture.com>,
<tel:+34609234>
```

The presence of the `P-Associated-URI` header lets the IMS client know which is/are the actual public IMS address(es) registered for that user. Public addresses can be used now to access IMS services. In particular, the PoC/XDMC client may use one of the listed addresses to initiate PoC sessions or to access XCAP documents. Observe that an agreement between the end user and the service provider must exist so that these addresses are provisioned into the IMS, and the user is aware of her assigned public IMS identity.

Observe that until the PoC/IMS handset does not register to the IMS, it cannot know which IMS address the user has been assigned, since the temporary addresses calculated based on the IMSI are only used for registration purposes. This configuration actually poses a challenge, in terms of correctly managing IMS, PoC and XCAP addresses in a consistent way. The issue is as follows:

- The XDMC must not use the temporary IMPU as described above: this can only be used for IMS registration.
- Additionally, if the IMS client has not registered to the IMS network yet (or if it has been administratively de-registered for some reason), the XDMC does not have any means to know what address to put into the `X-3GPP-Intended-Identity` of XCAP operations upon user request. In fact, if the user wishes to access a document in her folder, the XDMC does not even have enough information to insert the *XCAP User Identity* (XUI) in the XCAP URL of the request.

We can conclude that, with current 3GPP / OMA definitions, it is challenging for the XDMC to access the XDM service when the client is not registered to the IMS network. This is so because the XDMC does not have any means to determine a proper XCAP User Identity of the user. In order to overcome this problem, three (possibly many more) potential solutions can be thought of:

a) Perform IMS registration every time the user intends to access the XDM service. This requirement is actually quite strong, since one cannot assume that an XDMC is always co-located with an IMS client.
b) Let the user introduce her username as a configuration parameter of the XDMC. This option poses a serious security risk, as user access to service configuration is not linked to the availability of a valid SIM/USIM card, just to user controlled (or uncontrolled) configuration information.
c) Cache the values received in the P-Associated-URI SIP header after the first successful registration has occurred. This solution has similar drawbacks as point b.
d) Force that the MSISDN number is used as the basis to construct a TEL URI and insert it as the XUI for all XCAP requests and as the user address in all SIP requests. The issue with this approach is that a non-SIP native addressing scheme is used, which may lead to inefficient processing operations when routing SIP requests originated by the user.

It is not clear at the time of writing what technical solution will be implemented to overcome the problem described above. In any case, we believe that operators considering the launch of OMA PoC services over IMS based on SIM/ISIM access should consider the following elements:

• A mechanism to easily provision a Public User Identity for users that access the IMS service from a SIM equipped IMS handset should be implemented. Such mechanism should consider how to distribute such information in a friendly manner to subscribers, so that they can populate their corresponding *Addresses of Record* (AoR) [8] in business cards or mail signatures.
• A convention to enable XDMC access to the XDM service should be carefully designed to cope with the scenario in which the user is not registered into the IMS network.

In addition to assigning a valid IMS address during the registration process to users that access the PoC service from a SIM/USIM (no ISIM) equipped device, the authentication mechanism used for such access type is based on the so-called *Early IMS* model [34], rather than on the IMS Authentication and Key Agreement (AKA) mechanism defined for ISIM access.

Finally, it is important to understand that there are no specific addressing considerations to be discussed about the OMA Presence service. Presence information refers to presentities that are, by definition, IMS and PoC subscribers. Thus, a 'Presence address' concept does not exist, so there is no need to make any particular consideration about it.

5.4.2 Public Service Identities: PoC Group Addresses

IMS *Public Service Identities* (PSI) is used in the IMS environment to refer to resources stored or managed by Application Servers. As the reader may have imagined, PoC group identities are, from the IMS point of view, nothing more that PSIs. This assertion leads to several aspects that should be well understood by any service provider. We list these aspects below:

- PSIs are stored in the HSS, either as a unique identity or as a 'wildcarded' entry. The latter allows for the usage of templates, which help the HSS manage PSI's in a more effective way. Wildcarded PSI's are similar to URI templates described in [11].
- XDMCs must be provisioned with a URI template format as defined in [11] via DM operations, so that the resulting template is fully compliant with operator policies about PSI naming conventions.

Outside of the PoC domain, PSIs can be assigned to conferencing services, messaging services or any other type of resource managed by an Application Sever connected to the IMS via the ISC interface.

5.4.3 ENUM Service

As explained previously, the two addressing schemes used in IMS and PoC are the TEL URL and the SIP URI, the former being used in order to enable seamless interoperability with legacy networks and devices.

Observe, for example, that if a user has been assigned the following addresses:

- Public IMS User Identity 1 (PoC Identity 1): <sip:joe.doe@wirelessfuture.com>
- Public IMS User Identity 2 (PoC Identity 2): ⟨tel:+34-6054542⟩

He or she should be able to initiate and receive PoC calls from any of these addresses at any time (provided that both addresses are listed in the `P-Associated-URI` header upon successful registration).

As an example, another PoC user that only knows Joe's telephone number may try to initiate a PoC session with John by directly typing his MSISDN (6054542). The PoC client would then convert it into a suitable TEL URL (⟨tel:+34-6054542⟩) and initiate a PoC session with him.

However, routing of SIP messages within a SIP domain such as IMS is always based on the SIP URI schema: SIP proxies do not generally know how to route a SIP message having a TEL URL as a Request URI. When such case occurs for a telephony session, the BGCF is invoked to properly route the call, but when it comes to purely SIP-based services such as PoC, the BGCF is not an option.

Let's observe how this affects the PoC case. We will consider the Ad-hoc PoC session setup case, although the example is general to all PoC session types:

1. Joe starts a PoC session with Mary (sip:mary.dawson@wirelessfuture.com) and Bruce (tel:+34639349). In this case, the simplified SIP INVITE message sent by Joe's PoC client might look like:

```
INVITE sip:poc_conf-fact@wirelessfuture.com SIP/2.0
Route: <sip:p-cscf@wirelessfuture.com>
Route: <sip:s-cscf@wirelessfuture.com>
From: <sip:joe.doe@wirelessfuture.com>
To: <sip:poc_conf-fact@wirelessfuture.com>
(...)
<?xml version="1.0" encoding="UTF-8"?>
<resource-lists xmlns="urn:ietf:params:xml:ns:
resource-lists"
        xmlns:xsi="http://www.w3.org/2001/XMLSchema-
        instance">
  <list>
    <entry uri="sip:mary.dawson@wirelessfuture.com" />
    <entry uri="tel:+34639349" />
  </list>
</resource-lists>
```

2. The SIP INVITE message is routed towards Joe's home S-CSCF, which in turn proxies the request up to Joe's home PoC server. SIP routing here is based on Route headers and the request URI of the INVITE message (which points to the PoC conference factory URI associated to the PoC service). Up to this point, the fact that Joe has included a TEL URL in the list of callees has not affected SIP routing in any manner.

3. The PoC server (which, in this case behaves as a Controlling PoC function) needs to generate a new terminating SIP INVITE request for each recipient being invited to the Ad-hoc PoC session. The two new simplified INVITE messages will look like:

```
INVITE sip:mary.dawson@wirelessfuture.com SIP/2.0
From: <sip:joe.doe@wirelessfuture.com>
To: <sip:mary.dawson@wirelessfuture.com>
(...)
```

and:

```
INVITE tel:+34639349 SIP/2.0
From: <sip:joe.doe@wirelessfuture.com>
To: < tel:+34639349>
(...)
```

Observe that, in principle, the SIP network does not know how to route the second message (intended to reach Bruce). The fact that a TEL URL is used as the destination addressing scheme prevents any proxy (e.g. the S-CSCF) from implementing standard SIP routing procedures.

The mechanism defined in SIP and IMS networks to cope with routing SIP messages that contain TEL URL requests is called ENUM, as defined in [12]. The idea is that whenever a SIP node receives a SIP request containing a TEL URL in any of the fields which are required to perform a routing decision, an ENUM query may be performed. This query is based on the *Domain Name System* (DNS) service as available in most IP based networks. The response to the query consists on the corresponding SIP URI address associated to the user. In the above case, for example, a query requesting the SIP URI associated to ⟨tel:+34639349⟩ would return the SIP address <sip:bruce.williams@remoteoperator.com>.

We will not go into further details about the ENUM service. For our purpose it is enough to understand that ENUM is used to perform a translation from TEL URL to SIP URI schemes, and vice-versa. If a SIP URI address type exists for a user that is being addressed via her TEL URL, the corresponding mapping should be delivered as a result of a successful query to an ENUM server.

In general, S-CSCFs implement an ENUM interface to perform such queries. This is needed in order to let the S-CSCF handle numerous cases related to IMS usage (not necessarily related to the PoC service). Additionally, the PoC server may also implement an ENUM interface as an optional feature [13]. If the PoC server supports this feature, an operator or service provider may decide where ENUM queries associated to the PoC service are handled (i.e. IMS or PoC AS), and configure such policy when integrating the PoC service over IMS. An operator may decide that for consistency reasons all queries are performed at the S-CSCF, while other operators may consider that the PoC AS should be responsible for handling all ENUM queries related to the PoC service.

Regardless of the preferred approach, it is important to note that if the ENUM query is to be handled by the S-CSCF, this node is only able to implement this feature in the *terminating* leg. This is due to the following reasons:

- In case of Ad-hoc calls, the list of callees is contained in a resource list XML document which is carried as the payload of the SIP message. CSCF nodes do not have the means to interpret such content. Hence, they cannot determine whether a TEL URL is included in the list document.
- In case of Pre-arranged PoC group (dial-out) calls the SIP message only carries the identity associated to the PoC group. The actual list of members of the group is stored in the PoC XDMS. Consequently, the S-CSCF cannot determine *a priori* which is the list of callees, so it cannot perform any query when receiving the originating INVITE request.

Observe that the originating leg signalling is always routed towards the S-CSCF that serves the originating user, so no need to perform any translation arises before the request reaches the S-CSCF or the PoC AS.

For these reasons, there are only two trigger points where it makes sense to perform the ENUM query, either:

- a) the PoC AS performs the query once it has received the request from the originating user, and determined the list of callees for the session (e.g. by downloading the list of members of the PoC group from the PoC XDMS, or by reading the list of called users for Ad-hoc PoC calls),

  or:
- b) the S-CSCF performs the query for each terminating call leg in which a TEL URL address has been included in the Request URI of the message.

Once the ENUM query has been completed and the SIP URI has been received, the S-CSCF (or the PoC AS) may replace the contents of the Request URI or the To: header by the corresponding SIP address.

These two solutions are shown in the next two figures.

For sake of clarity, the split between Controlling and Participating PoC functions is not shown in Figure 5.5 and Figure 5.6. Only functional signalling flows are shown: the message names and format of the requests to the ENUM server are not actual ENUM/DNS message formats. For further details the reader is referred to [12].

Observe that for a given deployment it is desirable to configure only one of the two options, in order to keep functional consistence across network entities.

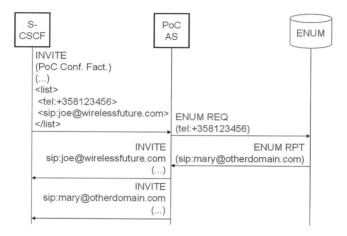

Figure 5.5   ENUM invocation from the PoC AS

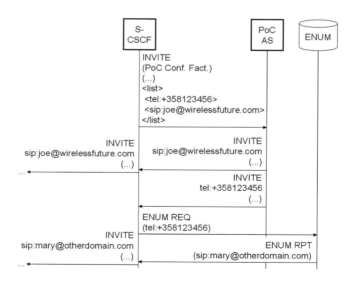

Figure 5.6   ENUM invocation from the S-CSCF at the terminating leg

5.4.4 PoC Group Identities and IMS Public Service Identities

An important IMS concept defined in [10] is the *Public Service Identity* (PSI). A PSI is an identity that points to an IMS Application Server, or a resource managed by an AS (such as the PoC server). This means that the requests that target a PSI are routed towards the relevant AS.

In order to properly manage user and service identities (IMPUs and PSIs) at the IMS domain and SIP messages containing them, network nodes (e.g. S-CSCF, HSS) may need to have a *hint* to distinguish between user identity addresses and service identity addresses. For efficiency reasons, this should be done 'at a glance', without the need to query a database (e.g. the HSS). For example, the S-CSCF may need to determine whether a request targets an IMS user (an IMPU) or a service (a PSI) in order to decide the best way to process such request:

1. If a request targets an IMPU, the S-CSCF needs to query the associated Terminating *initial Filter Criteria* (iFC, refer to section 5.5.1 for further details) to decide how to route the request
2. If a request targets a PSI, the S-CSCF may deliver the request directly to the AS hosting the identity.

Consequently, it is a good practice (probably, a need) for the operator to define a convention that lets S-CSCFs and HSSs easily distinguish PSIs from IMPUs. Consequently, we expect that PSIs and IMPUs will follow different structures, when being created by the end user, by a customer care agent or by any other entity provisioning user or service identities. Consider the following two sample addresses:

- Public IMS User Identity: <sip:joe.doe@wirelessfuture.com>
- Public Service Identity: <sip:Chatgroup&joe.doe@wirelessfuture.com>

If the HSS/S-CSCF has been provisioned with the rule that PSIs may be of the format: ⟨token⟩'&'⟨username⟩'@'⟨domain⟩, all nodes would be able to distinguish between SIP URIs pointing to IMPUs from those pointing to PSIs. In the above case, an address of the format ⟨username⟩'@'⟨domain⟩ (provided that the '&' character is not allowed for the username part of a SIP URI) would be correctly identified by the S-CSCF as a user identity.

In order to allow such differentiation among PSIs and other types of addresses, the wildcarded PSI concept has been defined in IMS [10]. A wildcarded PSI defines a template, which gives a hint to IMS nodes to determine how PSIs are formed. When a new PSI is created (by a user, a service or a provisioning agent) the PSI wildcard is used to ensure that the final name complies PSI policies defined by the operator.

In case of the PoC service, PoC group identities are considered IMS PSIs. Thus, whenever a PoC user initiates a Pre-arranged or Chat PoC group session, the identity inserted in the Request URI of the initial SIP request (e.g. SIP INVITE) is actually a PSI, from the IMS point of view.

PoC groups are owned and created by PoC users. The reader may remember that the XDMC is the entity actually talking to the PoC XDMS (via Aggregation Proxy) to create and manage XML documents (such as PoC groups). Hence, the XDMC is the entity responsible for creating PoC groups, according to the group naming convention that may be applicable in a given domain.

Table 5.1 Example usage of the URI Template mechanism in the XDMC

URI Template	Example URI generated from template
Sip:<id>@wirelessfuture.com	sip:mygroup@wirelessfuture.com
Sip:<id>_<user>@wirelessfuture.com	sip:mygroup_joe.doe@wirelessfuture.com
Sip:<id>&<user>@wirelessfuture.com	sip:mygroup&joe.doe@wirelessfuture.com
<xui>;poc-group=<id>	sip:joe.doe@wirelessfuture.com;poc-group=mygroup

As we have seen, wildcarded PSIs are used to define the format that PSIs should follow in an IMS domain. When translating this concept to the PoC case, the XDM framework has defined the 'URI template' concept [11]. The idea is that the XDMC be provisioned with a parameter that specifies the format that group identities as created by the user should have (whenever the user intends to create group documents in the future). Thus, when a user intends to create a new PoC group, the XDMC has a hint of what acceptable format the group name should have. By combining this URI template parameter with the XDMC GUI, the user may be prompted to introduce a 'token' (e.g. the 'group name') and the XDMC may build up the actual PoC group identity by appending the necessary elements to the token, as defined in the URI template.

For example, the user may introduce a string such as 'mygolfbuddies', while the actual PoC identity of the group may be 'sip:mygolfbuddies&joe.doe@wirelessfurure.com'.

The elements that can be used to construct a URI template are described in [11]. Table 5.1 provides some example formats of different URI templates and corresponding sample group identities.

In Table 5.1 the ⟨xui⟩ tag is replaced by a valid *XCAP User Identity* (which is expected to be equal to a valid PoC identity). The ⟨id⟩ tag is replaced by a unique XDMC-generated identifier (e.g. a group name as typed by the end user). The ⟨user⟩ tag is to be replaced by the user part of a valid XUI / PoC address.

Observe that the URI template is stored at the XDMC (via provisioning mechanisms). Hence, it is reused to build up all SIP identities that can be created from the XDMC, including Pre-arranged and Chat PoC groups, Presence subscription lists and any other future service requiring creation of groups or service identities that require SIP addressing for their correct operation.

5.5 IMS Connectivity

5.5.1 Filter Criteria: Triggering the PoC Service from the S-CSCF

In section 5.2 we have described the S-CSCF as the IMS node that provides services to a particular IMS user. In particular, the S-CSCF is responsible for routing SIP requests between IMS end users and their associated services, such as PoC, Presence or Messaging.

As more and more SIP-based services are deployed over IMS, the S-CSCF must be able to classify incoming SIP traffic and determine which AS should serve each incoming request. As an example, the S-CSCF should treat differently a SIP MESSAGE request containing a PoC Group Advertisement message, than a SIP MESSAGE request containing an instant message. The first one must be routed towards the PoC server, while the second one should be probably managed by a messaging server (e.g. the OMA Instant Messaging Server based on OMA SIMPLE Messaging [3]). Additionally, there are specific SIP messages that can be directly served by the S-CSCF itself (e.g. a SIP INVITE message establishing a 1-to-1 VoIP session).

The S-CSCF has to dynamically determine the 'next hop' in the chain of SIP nodes serving a SIP operation. Observe that, in general, this 'next hop' dynamic routing is only required for the initial request (e.g. INVITE, SUBSCRIBE, . . .) of a SIP dialog, because any subsequent message carried within an already initiated dialog will follow standard SIP routing principles to reach the intended recipient (e.g. based on Request URI, Via, Route, Record-Route, . . . headers). For dialog-less transactions (e.g. MESSAGE) the same concept applies only to the request path, as the response path also follows standard SIP routing.

3GPP IMS and 3GPP2 MMD have defined the *Filter Criteria* (FC) concept to cope with routing of SIP messages between the S-CSCF and ASs. 3GPP/3GPP2 realised that this concept only applies to *initial* SIP requests (e.g. INVITE, SUBSCRIBE, MESSAGE), that is: requests that initiate a SIP dialog or dialog less requests. Hence, the more popular *Initial Filter Criteria* (iFC) concept has been established.

The iFC is an integral part of the information related to an IMS subscription: the *User Profile*. In fact, the iFC virtually contains most subscription information provisioned in the HSS for a given IMS user. The iFC is downloaded by the S-CSCF during the IMS registration stage, and is stored in the S-CSCF for the duration of the registration. Additionally, the iFC can be downloaded at any time upon request by the S-CSCF. The iFC is an XML-encoded document that virtually provides all necessary routing information to the S-CSCF, so that it is able to uniquely determine how to route any potential incoming SIP message initiated or terminated by the user to which the iFC relates.

iFC settings are based on the asserted identity of the IMS user. That is: each IMS user identity has an iFC associated to it as part of the IMS provisioning information. The S-CSCF takes the value located in the trusted `P-Asserted-Identity` SIP header present in the SIP message, as the index to decide which user's iFC should be applied. The `P-Asserted-Identity` header is inserted by the P-CSCF, and contains one of the public identities previously registered with the REGISTER transaction. Further details on `P-Asserted-Identity` are provided in chapter 6. For the time being, it is worth understanding that it contains a public user identity that is trusted within the IMS domain (as opposite to the identity carried in the From header, which are not trusted by the IMS).

It is worth noting that filter criteria are applied by the S-CSCF both in the originating and in the terminating call leg. The originating case (*Originating Filter Criteria*) is triggered whenever the IMS user initiates an operation (e.g. PoC Session, Group Advertisement, IPA). In this case, SIP messages typically arrive to the S-CSCF via the P-CSCF.

The *Terminating Filter Criteria* are required to serve a user whenever a SIP transaction having that user as the recipient reaches the S-CSCF. There are two typical cases in which such filter criteria are required, namely:

a) Transactions arriving from an interconnection with another IMS domain (that is: transactions arriving to the S-CSCF via the I-CSCF or directly from a SIP node located in the other domain)

b) Transactions generated by an application. For example, the PoC AS issues a SIP INVITE message to each invited user when setting up an Ad-hoc session. Every INVITE message sent in the terminating legs triggers a routing decision at the S-CSCF based on the *Terminating Filter Criteria* of each called party.

The existence of *Originating* and *Terminating Filter Criteria* should not confuse the reader, in the sense that both cases are *Initial Filter Criteria*, applied only to initial SIP requests. All subsequent signalling messages within a dialog are treated according to standard SIP routing principles.

Let us present an example to illustrate how Initial Filter Criteria works:

1. Imagine that a PoC/IMS user (sip:joe.doe@wirelessfuture.com) registers an IMS identity upon start up of her IMS/PoC enabled device. Once the user is successfully authenticated and registered, the S-CSCF downloads from the S-CSCF the User Profile associated to user identity sip:joe.doe@operator.com

2. At a later point in time, Joe decides to start an Ad-hoc PoC session inviting a couple of contacts: Mary and Alice. It therefore generates a SIP INVITE message, which will reach the S-CSCF (the message is routed towards the S-CSCF of the calling user). The message would look like:

```
INVITE sip:poc-ConferenceFactory@wirelessfuture.com
SIP/2.0
Via: SIP/2.0/UDP device56.operator.com;branch=
   z9hG4bK77asdhds
Max-Forwards: 70
From: <sip:joe.doe@wirelessfuture.com>
To: <sip:poc-ConferenceFactory@wirelessfuture.com>
P-Asserted-Identity: "Joe Doe" <sip:joe.doe@
   wirelessfuture.com>
Call-ID: a84b4c76e66710@operator.com
CSeq: 314159 INVITE
Accept-Contact: *;+g.poc.talkburst; require;explicit
Contact: <sip:joe.doe@operator.com>;+g.poc.talkburst
Supported: timer
User-Agent: PoC-Client/OMA1.0 estreet_pocclient/v0.0.1
Session-Expires: 1800;refresher=uac
Allow: INVITE, ACK, CANCEL, BYE, REFER
P-Alerting-Mode: MAO
Require: recipient-list-invite
Content-Type: multipart/mixed
Content-Length: (...)
(...)
```

The content sections of the SIP message have been omitted, as they do not influence the iFC discussion (the above message would typically carry an SDP media description and a URI list containing the list of participants invited to the PoC Ad-hoc session).

It is easy to observe that in the above message there are certain elements that are specific of the PoC service. These elements have been remarked in bold. Effectively, a careful study of such INVITE message would lead us to the conclusion that it is a PoC specific SIP message. The three aspects that would lead us to this conclusion are:

a) Usage of the PoC *feature tag* string (`+g.poc.talkburst`) in the Contact and Accept-Contact SIP headers
b) Usage of the string `PoC-Client/OMA1.0` value in the User-Agent header.
c) Usage of the PoC conference factory SIP URI. For each given domain, there is a well-defined PoC conference factory, which is provisioned in PoC handsets, and is well known through the provisioning chain of the operator.

Therefore, we could construct a rule such as the following:

Whenever the S-CSCF receives an initial request (e.g. INVITE) from a valid IMS user, that . . .
- *. . . points to the PoC conference factory SIP URI (AND/OR)*
- *. . . the request contains the PoC feature tag (+g.poc.talkburst) in the* Accept-Contact *header (AND/OR)*
- *. . . the request contains a* User-Agent *header with the 'PoC-Client/OMA1.0' string in it,*
Then proxy the request further to the PoC AS".

The above rule summarizes the meaning of the iFC concept defined for the 3GPP IMS / 3GPP2 MMD. Although iFC are defined using an XML-encoded format, the goal of the iFC is exactly the same as the text in the paragraph in italic (i.e. defining a dynamic SIP routing rule). Each user has an associated iFC stored in the HSS. The iFC document covers all potential messages that can be received in the S-CSCF, so that in all cases the S-CSCF is able to perform a unique and correct routing decision based on the received message characteristics and iFC information.

In general, although OMA PoC clearly specifies the format of the User-Agent header, it is not expected that iFC rules will be based on such information, as it is mainly used for informational and debugging purposes. Additionally, since the PoC conference factory is only used in PoC Ad-hoc sessions, it is not reliable to base iFC on such information (because it is not present in Pre-arranged or Chat PoC group sessions). As a matter of fact, PoC iFC is generally based on the presence of the PoC feature tag ('+g.poc.talkburst') in the Accept-Contact header.

As indicated, Terminating Filter Criteria is applied by the S-CSCF that terminates an initial SIP request. This is:

- When the PoC server has received a session initiation request from a calling user, it has determined that there are one or more called parties to be alerted, and has generated a set of SIP INVITE requests targeted to the invited participants. In such case, the originating

PoC server will send these INVITE messages towards the S-CSCF through the ISC interface. The S-CSCF will then perform a routing decision applying the iFC rules for each
invited party.

Observe that one or more invited parties may reside in remote IMS domains. In this
case, the iFC may include a default rule indicating how to route messages to a
remote domain. Typically, the iFC will contain the *Fully Qualified Domain Name*
(FQDN) of the I-CSCF at the other network. This is: a domain name is provisioned as
the entry point to the remote network. The S-CSCF must convert that name into a suitable
IP address through a DNS query, before proxying the message towards the intended
recipient.

- When the PoC call has been initiated in a remote domain. In this case PoC messages are
  sent from the Controlling PoC server in the remote domain towards the terminating user's
  IMS network. Typically, messages will traverse the I-CSCF and reach the S-CSCF serving
  the terminating user. At this point the Terminating Filter Criteria will apply.

In Figure 5.7 we can see an example for both originating and terminating iFC.

- Originating iFC example. When a user initiates a PoC session (i1), the SIP INVITE
  message is routed from the P-CSCF to the S-CSCF (i2). The S-CSCF has previously
  downloaded the iFC associated to the calling user (possibly during the registration
  process). Information contained in the iFC file indicates that the SIP message should
  be routed to the PoC AS, since it contains the PoC feature tag in the `Accept-Contact`
  header (i3).
- Terminating iFC example. When a session is started in a remote domain, the originating
  request reaches the PoC AS in the remote network (the SIP request is routed within
  the remote network following *Originating* iFC routing). The PoC AS in the originating
  domain determines that a called PoC user is located in another domain (e.g. wirelessfuture.
  com). In such case, the PoC AS sends a SIP INVITE message to the S-CSCF via

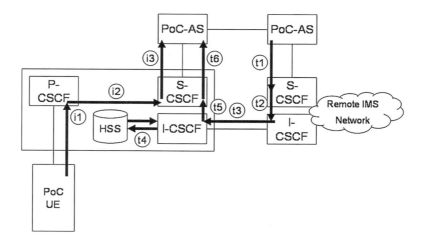

Figure 5.7 Originating and Terminating Filter Criteria Example

ISC interface (t1), and it includes the address of the called user in the Request URI of the newly generated SIP INVITE request. The S-CSCF detects that the message should be routed to another IMS domain, so it proxies the message further to the local I-CSCF (t2). The I-CSCF is a SIP proxy capable of routing messages to/from remote domains. It determines the address of wirelessfuture.com's I-CSCF and forwards there the SIP message (t3). The I-CSCF queries the HSS to determine which S-CSCF is serving the called user (t4). As a result of the query, the I-CSCF proxies the message to the S-CSCF (t5), which applies terminating iFC criteria and forwards the message to the PoC AS at wirelessfuture.com (t6). Observe that in this case the S-CSCF has taken the address in the Request URI as the index to determine what iFC to apply. This is opposed to the Originating Filter Criteria, where iFC is applied based on the P-Asserted-Identity header value. In both cases, however, the presence of the PoC feature tag is the key decision element to trigger delivery of the SIP message to the PoC server.

Once the iFC concept has been presented, it is possible to build up a matrix of iFC triggers that should let the S-CSCF determine when to proxy an incoming SIP message towards the PoC application. First, let us consider the list of initial SIP messages that should be considered for such matrix:

• SIP INVITE messages initiating a PoC session
• SIP REFER messages used to add participants to a session or to start a PoC when a Pre-established session is in use
• SIP MESSAGE requests carrying IPA or Group Advertisement messages
• SIP PUBLISH requests carrying PoC settings information
• SIP SUBSCRIBE requests requesting subscription to the conferencing event package.

Taking this set of PoC-related SIP messages we can easily construct a matrix that summarizes the iFC settings for a generic IMS/PoC subscriber. Table 5.2 has been encoded using pseudo-code logic commands. The actual format of an iFC XML document is described in [14][15].

Observe that since the iFC is a generic concept, it should include routing information for any initial SIP message that can be received at the S-CSCF. Consequently, iFC should define how to route SIP messages related to any service, such as PoC, Presence, conferencing, messaging, . . . If we consider Presence-related iFC there are only two types of initial SIP messages to be included in the iFC configuration, namely:

• SIP PUBLISH messages sent by Presence Sources
• SIP SUBSCRIBE messages sent by Watchers (when subscribing to Presence information about Presentities) and SIP SUBSCRIBE messages sent by Presence Sources when subscribing to the winfo event package – to receive a notification each time a new Watcher intends to subscribe to Presence information.

The Presence iFC configuration might look like Table 5.3.

We do not present XDM iFC configurations, since they are analogous to the concepts already presented. Additionally, SIP-based interfaces are optional for all XDMS defined by

Table 5.2 iFC used in the IMS to route PoC-related initial SIP messages

Originating Filter Criteria	Terminating Filter Criteria
Originating Filter Criteria is indexed on the P-Asserted-Identity header. CASE method='INVITE' AND header= 'Accept-Contact'='+g.poc.talkburst' THEN: ROUTE request to PoC Server Originating Port Address CASE method='MESSAGE' AND header='Accept-Contact'='+g.poc.talkburst' THEN: ROUTE request to PoC Server Originating Port Address CASE method='MESSAGE' AND header= 'Accept-Contact'='+g.poc.groupad' THEN: ROUTE request to PoC Server Originating Port Address CASE method='SUBSCRIBE' AND header='Accept-Contact'='+g.poc.talkburst' THEN: ROUTE request to PoC Server Originating Port Address CASE method='PUBLISH' AND header='Accept-Contact'='+g.poc.talkburst' THEN: ROUTE request to PoC Server Originating Port Address. CASE method='REFER' AND header='Accept-Contact'='+g.poc.talkburst' THEN: ROUTE request to the specified PoC Server Originating Port Address.	Terminating Filter Criteria is indexed based on the SIP Request-URI. CASE method='INVITE' AND header='Accept-Contact'='+g.poc.talkburst' THEN: ROUTE request PoC Server Terminating Port Address CASE method='MESSAGE' AND header='Accept-Contact'='+g.poc.talkburst' THEN: ROUTE request to PoC Server Terminating Port Address CASE method='MESSAGE' AND header='Accept-Contact'='+g.poc.groupad' THEN: ROUTE request to PoC Server Terminating Port Address CASE method='SUBSCRIBE' AND header='Accept-Contact'='+g.poc.talkburst' THEN: ROUTE request to PoC Server Terminating Port Address NOTE: Observe that neither REFER nor PUBLISH messages are received in the terminating case.

Table 5.3 Presence iFC configuration

Originating Filter Criteria	Terminating Filter Criteria
NOTE: Originating Filter Criteria is indexed on the P-Asserted-Identity header CASE method='SUBSCRIBE'AND header= 'event'='presence' AND header='Supported'='eventlist' THEN: ROUTE request to Resource List Server CASE method='SUBSCRIBE' AND header= 'event'='presence' AND NOT header='Supported'='eventlist' THEN: ROUTE request to Presence Server CASE method='SUBSCRIBE' AND header='event'='presence.winfo' THEN: ROUTE request to Presence Server CASE method='PUBLISH' AND header='Event'='presence' THEN: ROUTE request to Presence Server	NOTE: Terminating Filter Criteria is indexed based on the Request-URI. CASE method='SUBSCRIBE' AND header='event'='presence' THEN: ROUTE request to Presence Server CASE method='PUBLISH' AND header='Event'='presence' THEN: ROUTE request to Presence Server NOTE: It is assumed that in the terminating case all subscriptions are sent towards the Presence Server (i.e.: directly from the Watcher or back-end subscriptions sent by the Resource List Server).

OMA, and subscription to changes in XML documents is an optional client feature. Regardless, the reader may well infer how XDM iFC settings would look like, by inspecting the information presented until this point. In particular, subscription to PoC XDMS, Presence XDMS, RLS XDMS and Shared XDMS should handle delivery of SIP SUBSCRIBE messages requesting notification of changes in PoC groups, PoC rules, Presence policies, RLS resource lists and shared lists respectively.

5.5.2 PoC Network-Network Interface

At this stage it is clear that an integral part of PoC communications is the connection of different PoC domains, possibly served by different cellular operators. There is not only a technical requirement to connect PoC networks, but also a strong market demand: PoC take up and user adoption will only take place once cellular users are able to seamlessly PoC with their peers and colleagues. This is particularly true in case of residential (non-business) markets: as PoC becomes ubiquitously available in cellular handsets, users should perceive this service as a universal as regular circuit switched calling or text messaging. In order to achieve this goal, interconnection of PoC networks is a key milestone. This is to let users talk to each other, regardless of the cellular operator that provides the basic PoC service to each individual.

In order to enable this mid/long-term goal, OMA PoC already supports the basic tools to enable PoC interconnection. Although OMA does not include any particular section devoted to its specification, the set of features required to support PoC communications across domains is collectively known as OMA PoC NNI (Network-to-Network Interface). In this respect, IMS interconnection is a prerequisite to enable OMA PoC NNI. Effectively, interconnection of IMS domains enables SIP message routing between peer networks, as the basis to enable communication between PoC servers, being IMS Application Servers.

Figure 5.8 shows a high level overview of the PoC NNI architecture implemented over IMS. Observe that media and talk burst traffic flows directly between the PoC UE and its serving PoC AS, while SIP signalling traffic is proxied by 3GPP IMS, including interconnection of IMS (SIP) domains when needed.

Communication across PoC/IMS domains is basically a matter of performing proper SIP routing, and is largely based on ensuring that the proper DNS entries for SIP domains are used. When it comes to media and floor control routing, it is important to note that all POC-4 endpoint addresses and ports exposed by each PoC AS must be globally routable and reachable by any remote PoC network. This is similar to GPRS roaming requirements, where certain GPRS interfaces must be kept in the public IP addressing domain to enable proper interworking of services when users are roaming.

Finally, it is interesting to note that in a case where the IMS network provides services to both IPv4 and IPv6 enabled handsets, some interoperability issues may appear when interconnecting with IPv6-only IMS networks (IMS is in principle an IPv6-only domain, although the industry widely acknowledges that initial IMS deployments will take place over IPv4 as well [15]). In a case where two domains operating different IP versions are interconnected, the PoC AS should seamlessly support IPv4-to-IPv6 interworking, and vice-versa (dual-stack implementation). Another alternative is to use an IPv4-IPv6 Application Level Gateway and IP protocol translator [16]. This architectural implication may pose

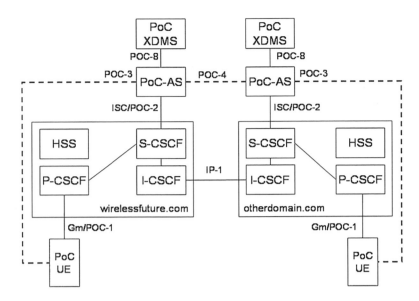

Figure 5.8   OMA PoC interconnection architecture

the requirement that the SDP implementation of the PoC AS must support proper siganlling of IPv6 addresses.

Observe that OMA NNI supports communication between the logical Participating and Controlling PoC functions. This means that each endpoint of an NNI interconnection behaves either as a Controlling or Participating PoC function. Since only one Controlling PoC function exists for each PoC session, the structure of OMA PoC interconnection is based on a 'star': all Participating PoC functions are connected to the same Controlling PoC function. This architecture supports the following features:

- Centralized floor control and media replication control point.
- Cascading structure: users may split across domains, allowing for an arbitrary large number of session participants (although scalability issues may arise if the Controlling PoC Function is not able to handle a large number of NNI connections and session participants).
- Basic SIP routing principles are used across all NNI connections.
- Depending on operator policies, the Participating PoC function may behave as a B2BUA or as a SIP Proxy when connecting to the Controlling PoC function.

Additional details about the Controlling and Participating split are presented in chapter 2. It is worth noting that, given the criteria used to decide which PoC AS behaves as a Controlling PoC function[4], it is potentially feasible that a PoC AS performing the Controlling PoC function is handling a PoC session involving several PoC servers in remote domains, while no user from the Controlling domain is actually participating in the session. This architectural implication may represent some technical (e.g. routing efficiency and signalling delay) and non-technical (e.g. charging implications) challenges, which may be addressed by OMA when specifying the evolution of the PoC standard towards version 2 of the enabler.

---

[4]E.g.: in case of Prearranged PoC group sessions, the PoC AS connected to the PoC XDMS that hosts a group performs the Controlling PoC function.

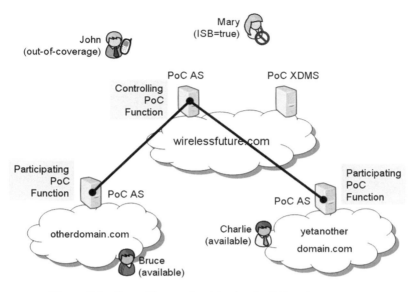

Figure 5.9   Controlling vs. Participating PoC Function use case

As an example of the above paradox, consider the following case:

1. John (john@wirelessfuture.com) creates a PoC group in his home PoC XDMS located at
   wirelessfuture.com domain. The group includes the following members:
   • Mary (sip:mary@wirelessfuture.com)
   • Bruce (sip:bruce@otherdomain.com)
   • Charlie (sip:charlie@yetanotherdomain.com)
2. John sends an OMA PoC Group Advertisement to all group members, and they store
   the address of the newly created group (e.g. sip:mycolleagues-john@wirelessfuture.
   com).
3. John switches off his PoC handset. Mary joins an important meeting and activates PoC
   ISB.
4. Charlie initiates a Pre-arranged PoC group call to sip:mycolleagues-john@wirelessfuture.
   com. Effectively, only Bruce accepts the session, so the PoC AS at wirelessfuture.com
   behaves as the Controlling PoC function, while only users accessing the group from
   otherdomain.com are involved in the session (in fact, the call actually becomes a 1-to-1
   session, although that does not represent any additional implication, apart from lower
   usage of radio and core network resources).

Finally, we present below a sample use case that shows how the overall NNI mechanism
based on IMS interconnection works. The example is based on Figure 5.10, and the
reader will be able to follow how iFC is invoked to trigger routing decisions at the
S-CSCF.

For the sake of clarity, some elements such as P-CSCF's or SIP temporary responses (1xx)
have been omitted:

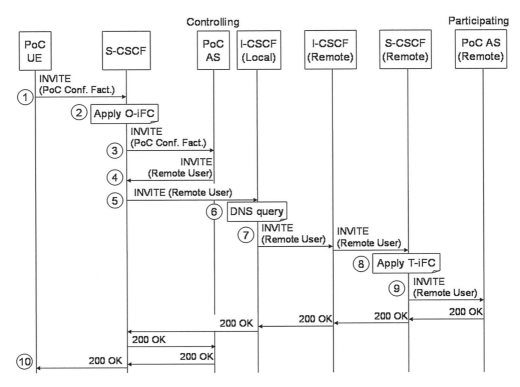

Figure 5.10   PoC session involving communication between two PoC (IMS) domains

1. Initially, the calling user initiates an Ad-hoc PoC session. The target URI of the SIP
    INVITE request is the PoC Conference Factory URI.
2. Eventually, the SIP INVITE message reaches the S-CSCF that serves the originating user.
    The S-CSCF searches the *Originating* iFC configuration for the calling user (as identified
    by the P-Asserted-Identity header inserted by the P-CSCF). iFC determines that the
    request must be routed to the PoC AS (the routing criterion is based on the presence
    of the PoC feature tag +g.poc.talkburst).
3. The SIP INVITE message is proxied further towards the PoC AS. The PoC AS retrieves
    the list of called users from the message (at this stage the PoC AS would possibly send
    a 1xx answer to the calling user).
4. The PoC AS determines that it needs to send another SIP INVITE message towards the
    intended called user, so it creates a new INVITE message (this is a new dialog initiated
    as a consequence of the incoming INVITE, hence the B2BUA nature of the PoC AS in
    this case).
5. The S-CSCF receives the newly initiated SIP INVITE request. In this case, the Request
    URI of the message points to the called PoC user. However, the S-CSCF detects that the
    Request URI belongs to a different domain, so it forwards the message towards a pre-
    configured I-CSCF that is capable of handling SIP messages intended to remote
    networks.

6. The I-CSCF performs a DNS query to obtain the address of the entry point to the remote domain (e.g. it queries the domain part of the Request URI included in the INVITE request). For example, if the call was targeted to sip:bruce@otherdomain.com, the I-CSCF looks for a SIP server associated to domain 'otherdomain.com'. The I-CSCF receives the address of its counterpart at otherdomain.com.

7. The I-CSCF at the originating domain proxies the SIP INVITE message to the I-CSCF at the remote domain. The I-CSCF queries the HSS and obtains the address of the S-CSCF that serves the called user at the remote domain.

8. The S-CSCF at the remote domain looks for a trigger entry in the iFC. Observe that in this case *Terminating* iFC apply, since the SIP message is received from a remote network (not from an originating user through the P-CSCF). Hence, the SIP address in the Request URI of the SIP INVITE message is used as the index to apply the most suitable iFC.

9. The request is finally delivered to the PoC AS at the remote domain that is in charge of serving sip:bruce@otherdomain.com).

10. In this particular example, Bruce has configured his automatic answer mode, so the remote PoC AS sends back a 200 OK response, which proxied back up to the originating user (traversing the remote PoC AS, the remote IMS network, the home IMS network, the home PoC AS and the home access network, including the P-CSCF).

The reader should, at this stage, be familiar with the main paradigms of PoC interconnection through IMS domains, and how interconnection and iFC concepts are combined together to enable PoC session setup in any case including intradomain and interdomain communications (which eventually may involve a large number of participants accessing the session from several PoC/IMS domains).

5.5.3 Third Party Registration and the Registration Event Package

In certain cases, SIP Application Servers may be interested in knowing the SIP/IP Core registration status of end users. This information may be useful in deciding how to serve an incoming request. The IMS supports two basic mechanisms to make applications aware of end user registration status. These two mechanisms are, namely: *Third Party Registration* and *Subscription to the Registration Event Package*. Since these features may be interesting from the PoC and Presence perspective, we will provide a brief description of these capabilities below.

Third Party Registration [14] is a simple mechanism implemented by the S-CSCF. It basically consists on relaying the authorized REGISTER message sent by the IMS client towards Application Servers within the trusted domain (i.e. typically, ASs co-located in the same domain as the S-CSCF). A sample signalling flow is shown in Figure 5.11.

1. The regular registration-challenge process takes place between the IMS client (e.g. the PoC UE) and the S-CSCF.

2. Upon successful registration, the S-CSCF relays the REGISTER message to Application Servers within the trusted domain that are interested in receiving registration status about the user (e.g. the PoC AS). The REGISTER message carries the explicitly register Public User Identity (IMPU) in the To: header.

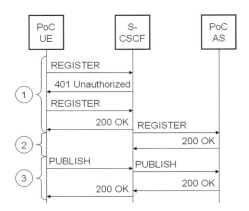

Figure 5.11   Third Party Registration mechanism

The S-CSCF may be provisioned with the list of ASs for which Third Party Registration
messages are relevant (i.e. some ASs may not need to receive these messages for their
regular operation, so there is no need to perform Third Party Registration for all servers
in the trusted domain).
3. Subsequently, the PoC client may PUBLISH the PoC service settings, in order to be reachable for incoming PoC sessions.

Subscription to the Registration Event Package is defined in [27]. It is a slightly more
complete mechanism than Third Party Registration. The idea in [27] is that network
entities interested in receiving detailed registration information about a user may SUB-
SCRIBE to their registration status (using the `Event:  reg` header). The S-CSCF is the
entity handling these subscriptions. As a result, the S-CSCF sends notification messages
to the subscribing entity; such messages contain detailed registration information about
the user, such as:

• Explicitly and implicitly registered user identity/ies.
• Registration status for each identity (e.g. active, terminated, . . .).

Figure 5.12 outlines how this mechanism works.

1. The PoC AS interested in receiving detailed registration information about a user sends a
   SIP SUBSCRIBE message to the S-CSCF, including the `Event:  reg` header. The S-
   CSCF notifies the PoC AS indicating that there is no registered identity at the moment.
2. The user switches on the phone and the PoC client successfully registers to the IMS
   core.
3. Upon successful registration, the S-CSCF sends a SIP NOTIFY message to the PoC AS,
   providing the list of explicitly (sip:joe.doe@wirelessfuture.com) and implicitly (tel:
   +34123456789) registered identities.
4. Publication of PoC settings follows the regular process as described in [9].

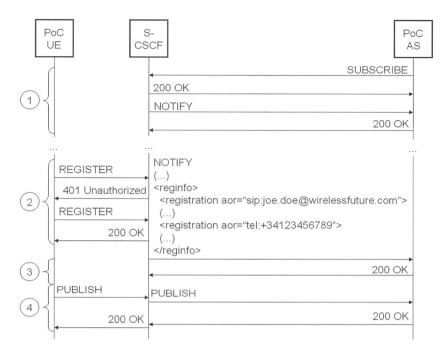

Figure 5.12   Subscription to the Registration Event Package

Observe that when detailed registration information is needed, [9] can be used, while for simple registration status Third Party Registration can be used. In general, there is no need to use both mechanisms simultaneously.

In general, there is no need for the PoC AS to support any of these features. Effectively, the PoC AS uses the PUBLISH message as the means to determine PoC User availability and answer mode settings. Regardless, there are several scenarios where these features may be of interest. To name a few:

- The PoC Server may use Third Party Registration information and PoC settings as published by the user in order to publish Presence information on behalf of the end user. This option can be used to save radio network resources and to ease client implementation. More details about this option are presented in section 4.6.2.
- Alternatively, the Presence server may also receive registration information, in order to fill up the IMS registration Presence fields, as described in [28].
- The PoC AS may also use registration information received using the Registration event package to obtain a detailed list of registered identities. This feature can be used to retrieve available SIP URI and TEL URI addresses. This list of addresses can be used when TEL URI to SIP URI conversions are required, and may help to avoid the need to implement an ENUM interface at the PoC server.

5.6 Charging PoC Services with IMS

In general, every service provider considering the launch or already operating a PoC service is interested in charging its customers. Effectively, operators deliver new services either to increase ARPU, decrease churn, differentiate . . . In summary: for one or another form of commercial reasons.

As other data services launched over cellular networks, such as WAP, text and picture messaging, or streaming and mobile-TV services, operators and service providers are interested in being able to charge and bill their PoC subscribers in one way or another.

It is first important to note the difference between two related concepts: *charging* and *billing*. *Charging* refers to the process run between network entities and the back-end systems, to obtain information about a particular event, collect such information, format it according to operator's requirements and evaluate it so that it is possible to determine (rate) the value of such event. In contrast, *Billing* refers to the actual generation of a *bill* (based on the information that is obtained during the charging process), that requires subscriber payment [19]. These concepts are further detailed below.

It should be noted that traditionally charging systems have been object of little attention from the standardization point of view. The fact that charging operations only involve network and back-end nodes run by an operator means that open standards are not as key as with user services. In the circuit switched domain, charging, billing and advanced Intelligent Network (IN) solutions have been based on a 'walled garden' approach, leaving little space for open solutions.

While cellular networks initially inherited this type of solutions, 3GPP has progressively been expanding its scope in the charging environment, and finally within Release-5 / Release-6 a charging architecture for SIP based services, including online and offline charging capabilities was developed. If we take into account the aim for horizontalizing *Service Delivery Platforms* (SDP) with IMS, it makes full sense to include definition of a standardized charging architecture as another key step in this direction.

Although the Release-5 charging initiative [18] was the first step towards a standardized charging architecture for 3GPP networks, we will use Release-6 charging architecture [19, 20] as the basis for discussion in this chapter. Release-6 introduced a set of architectural and technical solutions that make it more flexible for service providers and operators, so it is reasonable to assume that operators deploying 3GPP based charging solutions will progressively converge towards Release-6 charging.

Figure 5.13 shows the high level 3GPP charging architecture, and shows the particular case of PoC charging as specified in [21].

In general, there are two different approaches to how charging is implemented for a given service, namely: *offline* charging and *online* charging. These two mechanisms are typically associated to the type of cellular user accessing the service, namely: *prepaid subscribers* vs. *postpaid* (contract) *subscribers*. Although offline and online charging do not necessarily apply to only one or the other type of subscription, it is worth explaining here the need for online and offline charging in the scope of pre-paid and post-paid subscribers.

A contract or post-paid subscriber typically has established a relationship with the service provider by means of a conract. These subscribers pay their bill at the end of a billing period (e.g. a month). Among other information, the bill contains a detailed report of all consumed services.

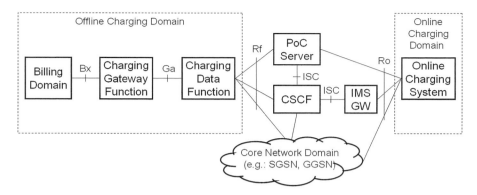

Figure 5.13   Release-6 charging architecture from the PoC perspective

Prepaid users pay their services in advance. Generally, a prepaid user purchases an amount of credit, which is then used every time a service is invoked from the prepaid account. When the credit is exhausted the user cannot access the service. The service provider may alert the prepaid user when low, zero or negative credit is available in the account.

These two paradigms build up the basis of the offline and online charging approaches defined in [19], and outlined in Figure 5.13. The general idea is that for postpaid subscribers the offline domain is used to charge and bill service usage, while for prepaid subscriber the *Online Charging System* (OCS) takes care of checking the user's credit and authorizing service usage. Observe that in the latter case, in general, there is no need to generate a bill for the end user, so no connection to the billing system is required. The main architectural elements of offline and online charging are described below.

5.6.1 Offline Charging Concepts

Offline charging is based on generation of so-called *Charging Data Records* (CDRs) (also, *Call Detail Record*). A CDR is a formatted collection of information about a chargeable event. For example, in case of circuit switched GSM calls, a CDR may contain (among other information) the following fields:

- Identity of the calling user, identity of the called user, call duration, date and time when the call was initiated, network identity where the calling user was located, network identity where the called user was located, . . .

In the 3GPP charging architecture, CDRs are generated by the *Charging Data Function* (CDF), which is the entry point to the offline charging domain. All network elements capable of generating charging information are connected to a CDF via the standard Rf interface. Rf is based on the Diameter protocol as specified by IETF [22]. Diameter is a modular protocol that supports several extensions for different types of purposes in the areas of authentication, authorization and accounting (AAA), and is widely used in the IMS domain. Diameter is used as a transport protocol over Rf interface to let network nodes carry charging information to the CDF. The CDF then generates a CDR for each information set received via Rf, and delivers the generated CDR to the *Charging Gateway Function* (CGF) [23].

In a 3GPP network there may be several network elements generating offline charging information in parallel. As an example, for a service such as PoC, several levels of CDRs may be generated by the CDF. A 1-1 PoC session between two PoC users may trigger charging reports being delivered by the SGSN, the GGSN, the P-CSCF, the S-CSCF and the PoC server. However, depending on operator policies, not all reports may be relevant.

5.6.2 Online Charging Concepts

A key requirement of the online charging system is to perform charging operations in a *real-time* manner. The idea of online charging is to query the user's credit before a service is effectively authorized, so that service usage is blocked if credit is not enough to fulfil the requested service.

In this case, service execution is closely linked with online charging operations (as opposed to the offline case, where service usage is merely *reported* to the charging function). It is important that interaction with the online charging domain is designed to be as effective and fast as possible, so that when the user has enough credit (which should be the general case) the user experience is not degraded due to credit check operations.

When a Network Element requires a credit check operation to authorize service usage, it queries the *Online Charging System* (OCS)[5] using the Ro reference point. Basically, all 3GPP core, IMS and service level nodes that need to invoke credit authorization operations will progressively incorporate the Ro functionality.

However, it is more effective to invoke the OCS from as few network elements as possible, so that only the information required to implement the selected billing model (and not more) is used. In the PoC case, for example, if the charging model is based on information available at the PoC server, it does not make sense that all underlying network nodes (GGSN, CSCF etc.) invoke credit check operations as well.

For each received credit operation, the OCS is responsible for [25]:

(a) Rating the received usage request (i.e. determining the monetary units that should be deducted from the user's account if the intended service usage is authorized).
(b) Performing reservation and account balance checks. In particular:
   (i) Determining whether the requested usage can be granted.
   (ii) Debiting (i.e. deducting) monetary units from the user's account, once a requested service has been effectively delivered to the end user.

The *Diameter Credit Control Application* (DCCA) [24] is the reference protocol for the Ro interface. DCCA natively supports credit check, credit reservation and credit granting operations, so it fits very naturally with the design requirements of Ro interface and online charging in general.

In the following paragraphs we will see how the charging concepts presented in this chapter apply to the PoC service.

---

[5] The Online Charging System is an entity that performs real-time credit control. Its functionality includes transaction handling, rating, online correlation and management of subscriber account balances. In essence, the OCS is responsible for performing credit check operations prior to service consumption, and credit deduction operations immediately after a service has been consumed. A key requirement of the OCS is to perform credit operations in real-time, so as to not affect the end user experience of the service.

5.6.3 PoC Charging Concepts

5.6.3.1 PoC Charging Models

PoC servers in commercial networks may progressively incorporate offline and online charging capabilities, that is: implement Rf and Ro interfaces respectively. PoC charging principles and signalling flows are specified in [21].

There are several chargeable events in the framework of the PoC service, for example: PoC session participation (i.e. exchange of talk bursts), sending of IPA and GA messages, or subscription events. We will focus our analysis on PoC session charging.

The following list provides a non-exhaustive outline of different approaches to PoC session charging. Depending on the actual business model that each operator selects, one or more options may be relevant.

• Flat-fee model. This is the easiest approach to PoC charging, in which users are allowed to access the service without any restriction during a certain period (e.g. a month). This model may include a cap limit, to avoid excessive malicious usage.
• Session duration charging. In this case, the user is charged based on the length of the PoC session that she initiates. This model emulates the widely adopted charging model for GSM circuit switched calls (where the calling party pays for the duration of the call).
• Talking duration model. This is a more finely tuned model in which the user effectively only pays for the time that she talks. Observe that this slight difference with the above represents an important technical difference: session based charging can be charged based on SIP signalling messages exchanged between parties, while talk duration requires processing TBCP messages in order to determine what charge to apply.
• Pay-per-push model. In this case, the user is charged a fixed amount of money every time she talks, regardless of the duration of the talk burst. Observe that PoC servers can revoke the floor of the talking user if he or she exceeds the maximum allowed talking time (T2 timer as defined in [17]).

Apart from the main paradigm of the billing model, several elements may lead to the definition of sub-models or tariff plans, or even the combination of the above options. Such modifications may be based on parameters such as:

1. The number of participants in a session. As an example, in the pay-per-push model, pushes may be charged differently based on whether the session type was 1-1 or 1-Many.
2. The location of the participants. For example, when a user is roaming into a remote location, roaming charges may apply to incoming sessions.
3. Group ownership charging. Even though a user may not participate into a group session, she may be charged for hosting a PoC session when two or more PoC users engage in a session with one of the Pre-arranged PoC groups owned by that user.
4. Charging based on identity of the participants. As an example, in a corporate tariff, sessions between users working for the same company may be charged at a lower tariff than calls to external users.
5. Time of the day. As an example, a consumer segment user may use a flat-fee model during low usage timeframes (e.g. evening), but a pay-per-push model during high usage time (e.g. morning).

So, even though the main charging model may be relatively easy to understand, combining it with the additional parameters described above may end up building up a relatively complex tariff plan. In any case, it is clear that the charging capabilities of the PoC server may need to provide a large set of information to the charging system (offline and online) in order to be able to apply any charging model that an operator may select. Obviously, in many cases a subset of such potentially large amount of information will be used. For example, the flat-fee model requires much less information than a pay-per-push model where the actual tariff depends on the identity of the session participants.

It is fair to mention that some charging models (e.g. session based charging or flat-fee) can be directly implemented at the IMS layer (S-CSCF). In this case, IMS charging mechanisms apply, thus the charging capability of the PoC server can be directly disabled. For more complex models it makes sense to assume that the PoC server is the only network entity interacting with the charging systems. This approach makes sense because the PoC server is the network element capable of processing more information about a PoC session (the GGSN only processes the IP layer, and the S-CSCF only understands SIP, and does not handle talk burst messages). In the following paragraphs, we will assume that all charging operations are triggered by the PoC server, thus Ro and Rf interfaces present at lower layers are not used for the PoC Service or discarded by the CGF (in case of offline charging).

5.6.3.2 Triggering PoC charging from the IMS

Before examining further how PoC offline and online charging work, it is interesting to see how PoC charging is triggered. We will present these concepts using Figure 5.14 below.

When the S-CSCF receives a SIP INVITE message to be delivered to the PoC AS (due to the presence of the PoC feature tag when applying iFC) the following events occur:

1. The S-CSCF determines that the originating user is a postpaid user, so offline charging should apply. In order to indicate this to the PoC server, the P-Charging-function header is inserted. This header is defined in [26]. It is used to provide Application Servers with the address of the relevant charging entity: the CDF for offline charging and the OCS in case of online charging. If both addresses are included, the AS should use Offline and Online Charging in parallel. The CDF address is provided in the CCF parameter, while the OCS address is indicated using the CFF parameter. This naming convention is for historical reasons, as [26] was being developed during Release-5 timeframe. Observe that in Figure 5.14 the S-CSCF mandates offline charging to be applied to this session.

2. A P-Charging-Vector header is included. This header is inserted by the P-CSCF and completed by the S-CSCF. It is used to provide several tokens and identifiers, which are useful for charging systems to correlate charging events triggered by different network elements, or charging events occurring at different times. As an example, an *IMS Charging Identifier* (ICID) is included in the ICID-value parameter. This serves the charging systems to relate all transactions related to the same IMS session. Other parameters are used to provide GGSN charging information, or to convey information about the (potentially) different domains involved in the session (due to roaming or interconnection calls). This information is used to support inter-operator CDR exchange operations.

3. Eventually, the SIP INVITE message reaches the PoC AS. It then determines that offline charging should apply, so it starts collecting information about the session and sends a

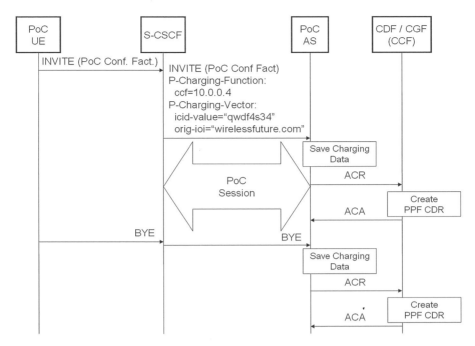

Figure 5.14 Configuring PoC charging from the S-CSCF. Offline Charging example

Diameter Accounting Request (ACR) to the CDF (CCF) indicating the start of a new service session.

4. We assume here that the PoC AS fulfils the Participating role, so the CDF creates an initial Participating PoC Function (PPF) CDR with the information received from the ACR message.

5. The session proceeds (e.g. participants exchange several talk bursts) and eventually it ends (e.g. all participants leave the session). When the PoC AS receives the final BYE message that effectively terminates the PoC Session a new ACR message is sent to the CDF, reporting all the necessary usage information (e.g. number of exchanged talk bursts, and/or session duration, and/or user talking time, and/or participants identity information).

6. When the final ACR message is received the CDF fills up the created PPF CDR with the new information (if relevant) and closes the CDR, which is now ready fore delivery to the CGF (unless CDF and CGF are co-located in the CCF). All ACR messages are acknowledged by the CDF with an ACcounting Answer (ACA) message.

Apart from usage of the `P-Charging-Function` header to indicate online/offline charging usage, the operator may statically configure such parameter in the Application Server. This is particularly interesting when one or both mechanisms apply to all customers within a domain.

5.6.3.3 PoC Offline Charging

PoC offline charging follows the mechanism described in the previous paragraph. In particular, all information exchanged between the PoC AS and the CDF is based on Diameter ACR and ACA messages. The ACR message is used to report to the CDF information about PoC sessions. Typically, an ACR message is sent at the beginning of the session and another one once the session is over (to report usage, so that the charging system can close the CDR and deliver it to the CGF. Additional interim CDRs may be generated based on a number of conditions. This is particularly useful to avoid loss of charging information if a PoC session lasts longer than a certain threshold.

It is even possible to generate a CDR for each sent talk burst (this is relevant in the pay-per-push model). While this approach is the safest one (if the server crashes, all past events will be charged properly, since they are communicated in near-real-time), it may overload the Rf interface (it is more efficient to report usage at the end of the session, where all talk bursts can be reported with a single ACR message). For each operator, there may be an optimal frequency of reporting that minimizes Rf interface traffic, while still ensuring that if the PoC AS crashes only a small set of charging information is lost.

5.6.3.4 PoC Online Charging

PoC online charging has heavier impact in service execution than offline charging. As an example, the charging model selected by an operator directly impacts how and when charging operations are triggered. The following cases illustrate this aspect:

- For the flat-fee model, it is enough to determine whether a user is authorized to invoke the service or not. Such a query can even be performed during PoC registration stage, so that it does not impact session setup delay when the user starts a PoC call. In this case, when the user publishes his PoC settings, they may be rejected or accepted based on the query to the OCS, and when a subsequent INVITE message is received the PoC AS does not need to query the OCS again (provided that the grant has not expired yet).
- For the session based model the PoC AS needs to perform credit check operations based on sent and received SIP messages.
- For the pay-per-push model, online charging requests are triggered each time the user requests the right to speak.

We can easily conclude that an off-the-shelf PoC server deployed in one or another service provider may need some further configuration to configure the trigger point for online charging operations. The following picture shows how a generic PoC online charging implementation could work.

1. When a PoC user requests initiation of a PoC session a SIP INVITE message reaches the PoC AS (eventually containing the relevant `P-Charging-Function` and `P-Charging-Vector` SIP headers).
2. The PoC AS determines that online charging be used and sends an initial *Credit Control Request* (CCR) DCCA message to the OCS. Remember that Ro interface uses the DCCA

protocol, which is intended to manage credit check operations, which makes it much more suitable and efficient than the Diameter Base application.

3. The OCS receives all information necessary to rate the requested service, compares it with the existing balance in the user's account and decides that the service can be authorized. It then blocks a fraction of the user's account and grants service usage (typically the amount of granted usage equals the value of the monetary units blocked in the account).

   (a) Blocked account credit cannot be used for any other service until the service that requested it (i.e. PoC) either consumes the credit or releases it.

   (b) The granted usage is reported by the OCS in the *Credit Control Answer* (CCA) message. Observe that, in general, the *Granted Service Units* (GSU) reported in the CCA are not monetary units, but service related units, such as: amount of pushes the user is allowed to send, number of seconds the user is allowed to speak).

4. The PoC AS receives the CCA message and starts managing the granted units. As an example, if the CCA indicated that the user is allowed to send 5 pushes, the PoC AS needs to count how many pushes the user sends during the session (by monitoring TBCP messages). Eventually, the PoC AS may need to query the OCS again, so that granted units are consumed and new units are requested. For such purpose a CCR (Update) message would be sent (not shown in Figure 5.15).

5. Finally, the user leaves the session and the PoC AS reports back to the OCS to indicate how many units were consumed and how many can be released. It is important to send the CCA (Terminate) message, so that the OCS can release blocked units from the user's account, as to avoid credit fragmentation.

Credit fragmentation is a general concept linked to the online charging paradigm. Consider for example a scenario where a prepaid user may access several services in parallel: he may try to send an MMS, while browsing through the Internet and participating in a PoC session. In this type of scenario, there may be several prepaid services blocking credit from the user's account in parallel. This situation may lead to blocking a user's access to a service while there is actually enough balance in the account. The problem is that most of the balance is blocked by several applications 'just in case'. The balance will only become available when one or more applications send the final CCR (Terminate) message reporting used and unused service units. As indicated, this is a universal online charging problem. There may be ways to mitigate it, by performing frequent queries to the OCS (and, thus, blocking small credit chunks). Hence, each operator may configure an optimum trade off between credit fragmentation risk and Ro interface efficiency, based on its charging models and credit management policies.

In Figure 5.15 we can detect that the way the PoC AS interacts with the OCS consists of performing an initial reservation upon service startup, authorizing the service if the reservation is successful and reporting back usage at the end of the session. This behaviour is quite common in the online charging environment (not only for PoC sessions), and is known as the *Session Charging with Unit Reservation* (SCUR) paradigm. PoC sessions are always charged following one or other form of the SCUR paradigm.

The other two main online charging paradigms are *Immediate Event Charging* (IEC) and *Event Charging with Unit Reservation* (ECUR). IEC is not applicable to PoC, while ECUR applies to charging of IPA and GA PoC events.

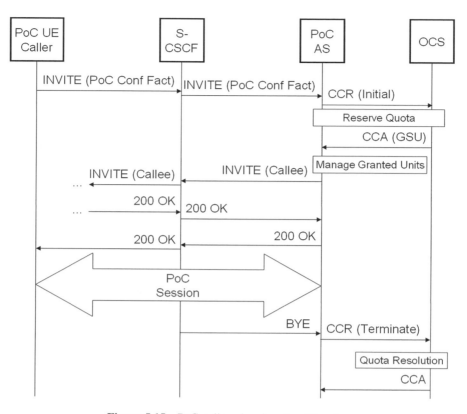

Figure 5.15   PoC online charging signalling flow

5.6.3.5 Overview of PoC Parameters in Charging Messages

We have seen that PoC charging may be based on offline and online mechanisms based on Diameter or DCCA respectively. Regardless of the mechanism actually used, Diameter is a largely extensible protocol which can carry virtually any information. This capability is used by applications (such as DCCA) to extend the basic functionality.

Diameter is extended by adding new *AVP*'s to the core specification. Each AVP is an *Attribute-Value-Pair* structure. Actually, each extension carries a definition of what the extension is (the *Attribute*) and what value is carried in the extension (the *Value*). Using this approach, a large number of new parameters have been added. Some of these parameters are relevant from the PoC service point of view, since they constitute the basic building blocks to let an operator define the charging models around their PoC service. In Table 5.4, we list several AVPs as defined by 3GPP which are carried within Diameter messages. In all cases, they can be used for offline charging (directly carried over Diameter Base) or for online charging (over DCCA) [21].

Observe that Table 5.4 does not contain information related to actual service usage (e.g. number of talk bursts, authorized talking minutes). This is because the actual usage is defined in Diameter and DCCA as the abstract Service Units concept. For each application,

Table 5.4   Non-exhaustive list of PoC related Diameter AVPs

AVP Name	AVP Description
Event Type	Defines the SIP method that triggered the request
IMS Charging Identifier	ICID value useful to correlate all charging information related to the same session
Calling Party Address	PoC address of the calling user
Called Party Address	In the PoC case it may contain the PoC Conference Factory URI or a PoC group identity
Timestamp	This may be used for time based billing models
PoC Server Role	Controlling vs. Participating
PoC Session Type	'1-1', 'Ad-hoc', 'Pre-arranged', 'Chat'
Number of Participants	Useful for charging models based on the number of participants
List of Participants	This is relevant if the actual charge may be depend on the list of participants in the PoC session (e.g. 'in-company calls for free').
PoC Session-Id	Identity of the PoC session as assigned by the PoC server
PoC Group Name	PoC group identity. Only included if the PoC session type is 'Pre-arranged' or 'Chat'

different types of Service Units may make sense, but in any case, the amount of usage granted by the OCS is always reported in the general GSU (Granted Service Units) AVP.

At this stage, it is worth observing that every PoC deployment implementing the Ro interface may require a significant amount of customization. Such customization includes the following elements:

• Activating the proper charging trigger points within the standard PoC signalling flows, in order to support the charging model defined by the operator.
• Selecting the actual list of useful AVPs relevant for the selected charging model.
• Defining the actual meaning and usage of the GSU AVP in scope of the applicable charging model.

Observe that apart from PoC session participation, PoC events such as Call-Back requests (IPA) and Group Advertisements (GA) can be charged as well. An operator may decide to activate IPA and GA charging, or leave it free for PoC users. The latter case makes certain sense, since both IPA and GA promote service usage, so they indirectly increase operator revenue. Regardless, it is feasible to active offline and/or online charging for IPA and GA. When this option is used, a 'one-shot' charge is applied to each IPA or GA message (there is no concept of *session* associated to IPA and GA).

Finally, it is worth noting that, apart from PoC charging, service providers may intend to charge Presence and/or XDM traffic as well. However, since both Presence and XDM are more a horizontal service enabler than a real service themselves, there seems not to be a clear approach about how to charge them, or even whether to charge them at all. In the particular case of Presence, service providers may decide to charge a fixed small fee for the service (flat-fee model), or on providing basic Presence information for free, which results in a much simpler implementation of charging capabilities, when compared to complex models that may be used in the PoC case (e.g. pay-per-push).

5.7 Device Management

An important concern for service providers is to ensure that PoC capable handsets are properly provisioned, so that users do not need to go into complex configuration operations in order to access the service. In this section we provide a non-exhaustive summary of provisioning parameters that need to be in place in order for the PoC service to be reachable from a PoC enabled handset.

Table 5.5 outlines a list of provisioning parameters required for proper communications using GPRS, IMS, PoC and XCAP capabilities.

Required provisioning information may be configured in the PoC UE through several means, such as:

- Direct pre-provisioning when creating the production SW version of the handset before commercial release.
- Binary SMS delivery based on the configuration format defined by the handset vendor.
- OMA Device Management operations based on [29].

These operations need to be carried to configure GPRS access and some key PoC and XDM parameters (e.g. PoC conference factory URI).

In general, when accessing the IMS/PoC service from a cellular network, a PoC enabled handset establishes a data connection with the core network, known as a *Packet Data Protocol* (PDP)-Context. Such PDP-Context is a logical association maintained end-to-end between the PoC UE, the RAN and the Core 2.5G/3G network. PDP-Context configuration parameters need to be generally configured in the UE by means of provisioning.

Additionally, certain PoC parameters need to be provisioned as well, as indicated above. IMS parameters are generally obtained dynamically, such as the P-CSCF address (received during PDP-Context activation) or the activation of SigComp capabilities (which is managed by the P-CSCF during IMS registration). Finally, some session specific settings are naturally negotiated during PoC session setup via SDP means.

Proper PoC handset provisioning may not be a trivial task, particularly while different handset types may require different types of binary short messages being delivered to them. As OMA Device Management becomes the default solution, operators may be able to progressively converge towards a single provisioning interface for handset configuration. Although this scenario will only be feasible in the long term (as it requires PoC and DM widespread adoption), it is important to note in this section that it is indeed feasible to ensure that handsets are properly configured, by delivering to them all the relevant set of information required to configure the different access layers (GPRS, IMS, PoC, XDM).

Table 5.5   UE PoC, IMS, XCAP provisioning parameters

Parameter	Level	Type	Source of Information
GPRS access parameters: Access Point Name, DNS address, Authentication Data	GPRS	Static	UE Provisioning
QoS profile and PDP-Context management	GPRS	Static / Dynamic	UE and RAN/Core capabilities HSS provisioning Configuration negotiated during GPRS establishment
IMS / PoC public identity	IMS	Static	ISIM information, OR Notified during the registration process
IMS private identity	IMS	Static	SIM/USIM/ISIM Information
IMS authentication keys	IMS	Static	SIM/USIM/ISIM Information
Address of the SIP Core Entry Point (P-CSCF)	IMS	Dynamic	PDP-Context setup parameter, OR DHCPv6 procedures
SigComp usage	IMS	Dynamic	Configured by the P-CSCF
PoC Conference Factory URI	PoC	Static	UE Provisioning
AMR packetization mode	PoC	Static / Dynamic	UE capabilities, AND SDP negotiation
IP address and UDP port of the PoC server for media and floor control	PoC	Dynamic	SDP negotiation
URI Template for Service Identities	XCAP / PoC	Static	UE Provisioning
Exploder URI for IPA messages	PoC	Static	UE Provisioning
Simultaneous Sessions Support Preestablished Session Support	PoC	Static	UE Provisioning
Maximum Group Size	PoC	Static	UE Provisioning
T10, T11, T13 timers for media and floor control handling	PoC	Static	UE Provisioning
Notification that the PoC server publishes Presence on behalf of the user	PoC	Static	UE Provisioning
XCAP Root URI	XCAP	Static	UE Provisioning
XCAP identity (if needed)	XCAP	Static	ISIM Information, OR Notified during the registration process
XCAP password (if needed)	XCAP	Static	UE Provisioning
XCAP authentication method (3GPP Authentication vs. HTTP Digest)	XCAP	Static	UE Provisioning

5.8 Radio Access Network Parameters

This section will introduce the network parameters designed to optimize the PoC users' Quality of Service (QoS) when using E(GPRS) and WCDMA RAN. It is important the service provider allocates the adequate radio resources to PoC and other real-time IP based services to ensure a profitable return.

5.8.1 E(GPRS) and PoC

Push-to-Talk over Cellular (PoC) is implemented using IP as the transport over standard E(GPRS) networks. The E(GPRS) network provides a bearer service that includes mobility management and radio access procedures for the PoC service (Figure 5.16). To be reachable, the PoC subscribers should be in *always-on* PDP context mode.

5.8.2 Extended Uplink TBF Mode

As a real-time service, PoC benefits from the *Extended Uplink Temporary Block Flow* (TBF)[6] Mode features as it ensures a steady flow of packets in the uplink direction. Extended Uplink TBF is a feature of the E(GPRS) radio interface that lets a UE extend the inactivity transmission period without releasing the TBF. This saves radio capacity and prevents unwanted break.

The main characteristics of the Extended Uplink TBF Mode features are:

- Uplink direction can be prolonged as configured by the operator, typically for 1–2 seconds
- Unintended and unwanted breaks can be effectively prevented in the uplink direction during active periods
- During the extension period, an opposite direction TBF can be established more quickly using *Packet Associated Control Channel* (PACCH)

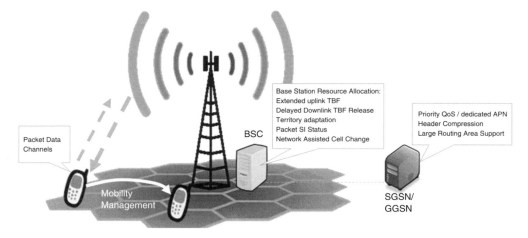

Figure 5.16   E(GPRS) and PoC

---

[6]The Temporary Block Flow is a concept associated with the E(GPRS) radio interface. Each time a device wishes to transmit data in the uplink direction, the network must make a temporary assignment of radio network resources to let the station transmit. We can associate a UE having an available TBF as a station that has 'permission' to transmit in the uplink direction. For real-time services such as PoC, having to request such permission every time a new RTP packet is to be transmitted may represent an important QoS issue. Rather, extended uplink TBF mode lets the handset keep the TBF for longer periods (even when no data is available for transmission) so that the real-time traffic does not experience variable delay over the wireless link.

- Some capacity has to be reserved for the extension period based on the configuration set by the operator
- Some radio capacity is saved by decreasing the number of random access procedures and the TBF establishment.

This feature needs to be supported both by the network and the terminal. The feature is specified in the GERAN Release 4 Feature Package 1 which is a collection of enhancing (E)GPRS features.

5.8.3 Delayed Downlink

The benefits of the Delayed Downlink Temporary Block Flow (TBF) Release feature are similar to those of the Extended Uplink TNF Mode feature. The Delayed Downlink TBF Release helps in establishing the uplink TBF faster and saving the system signalling resources. It is already implemented in most networks.

5.8.4 Priority Quality of Service (QoS)

The priority Quality of Service (QoS) feature makes it possible to differentiate the received QoS between services based on operator settings. The feature provides tools for arranging relatively higher throughputs to certain services and subscribers but it cannot arrange any absolute data rates.

It is recommended that PoC subscribers be allocated a specific APN, and that PoC users in that APN are allocated priority 1. For other services, such as WAP, Internet or e-mail, other APN(s) are allocated, and their priorities should be set lower than PoCs, at maximum, priority 2.

5.8.5 Header Compression

Headers form a significant portion of the payload. When the load is high, capacity can be gained by introducing header compression mechanisms in data transfer. This becomes more important when using IPv6, where the size of the IP header increases significantly.

In PoC speech data transfer, the packets consist of the following items:

- IETF AVT AMR frame (coded speech) – for example 110 octets
- IP header – 20 octets with IPv4 and 40 octets with IPv6
- UDP header – 8 octets
- RTP header – 12 octets
- SNDCP header – 4 octets
- LLC header – 6 octets

For PoC Degermark Header Compression (RFC 2507), RTP Header Compression (RFC 2507) and AMR framing into RTP packets (RFC 3267) can be used for reducing the load.

5.8.6 Territory Adaptation

The territory adaptation feature provides capacity and user gains if it is correctly tuned when the PoC service is launched in the network. Territory adaptation refers to the way Time Slots (TSLs) are divided between Circuit Switched (CS) and Packet Switched (PS) traffic. There are different types of TSLs in radio resource management:

- The dedicated GPRS TSLs are always dedicated to GPRS traffic, independent of the CS load

- The default GPRS TSLs are allocated for GPRS traffic as soon as the CS load is low enough, independent of the GPRS load

- The additional GPRS TSLs are allocated only temporarily for GPRS traffic when the CS load is low and if the GPRS load is high.

The CS part of the radio resource management continuously tracks the CS and PS load in the network and tries to hand over CS calls to other transceivers to free up space for PS load. If there are free time slots, the territory can be upgraded and the default time slots are available for GPRS traffic as soon as the CS traffic allows, independent of the PS load. Correspondingly, when there are too few idle time slots available for the new CS calls, or when the PS load fits into the default GPRS territory, the territory can be downgraded, after which there are more time slots available for the CS traffic.

5.8.7 E(GPRS) Mobility Management Improvements and PoC

For a real-time service like PoC it is important to minimise delays and the number of mobility events. For this purpose, there are features available, for example Packet SI Status and *Network Assisted Cell Change* (NACC).

5.8.7.1 Delays in Cell Reselection

Cell reselection is an area where end user perceived quality can be enhanced, even though the probability of a cell change taking place during an active speech burst is relatively small. The following mobility features are recommended for PoC:

- With the Packet SI Status feature, the cell reselection can be shortened significantly, as it is not necessary for the terminal to attempt to receive the complete set of System Information (SI) messages, but just the most important ones. The remaining messages left can be requested during data transfer in the target cell by using the Packet SI Status mechanism.
- With Network Assisted Cell Change (NACC) and Packet (P) SI Status supported, the cell reselection may be performed in even less than 700 ms. For PoC, this could be referred to as seamless mobility as the jitter buffers in the receiving terminal could be able to suppress the break and hide cell reselection. This feature requires support both from the network and the terminal.
- With Network Commanded Cell Reselection (NCCR), the network can guide particular terminals to particular cells. It can for example, guide EDGE-capable terminals to cells

supporting EDGE. NCCR is particularly useful when used together with NACC and Packet SI Status features.
- The Large Routing Area Support (LRAS) feature makes it possible to have larger routing areas, which minimizes the probability of Location Area Updates (LAU) and Routing Area Updates (RAU). This saves radio capacity and reduces unwanted breaks in the connection.
- 3GPP standards defined three different Network Modes of Operation (NMO I, II, and III) related to the coordination of Circuit Switched (CS) and Packet Switched (PS) services, NMO I is recommended to be used in the network as it:
  - Is the only way to guarantee reception of both CS and PS paging
  - Reduces latency in the location and routing area update process
  - Balances the paging load in the network.

5.8.8 WCDMA and PoC

The qualitative difference between WCDMA and E(GPRS) is the relatively constant bit rate in WCDMA (Figure 5.17). The benefits of WCDMA:
- Mobility; no interruptions at cell changes with soft handover
- Delays: Start to talk delay and Voice Through delay decreases after the initial setup of the radio bearer
- Capacity: wide bandwidth available make also larger PoC group sizes feasible
- Capacity: automatic/inherent non real-time and real-time trunking gain.

To utilise the WCDMA network capacity efficiently, it is important to allocate a Radio Access Bearer (RAB) with an appropriate bit rate and QoS class for PoC.

A low bit rate for the PoC service can be obtained by using a dedicated APN. The parameters in the QoS profile in the HLR for that particular APM should be set so that a low bit rate RAB is delivered from the network. By using a dedicated APN it is then possible control

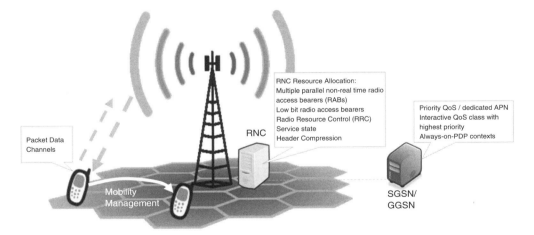

Figure 5.17   WCDMA & PoC

the bit rates that should be used for the PoC service in the radio access network as well as the QoS class/priority.

5.8.9 Multiple Parallel Non-Real-Time Radio Access Bearers

PoC should use a dedicated Access Point name (APN) with service specific bit rate and QoS settings. This means that a dedicated Radio Access Bearer (RAB) is used for the PoC service. Because it is also suggested to use an always-on PDP context with PoC, this RAB will be kept from the login to the logout from the PoC server.

5.8.10 Low Bit Rate Access Bearers

To enable large PoC groups and to utilise the radio resources efficiently, it is important that the PoC service is mapped to the low bit rate Radio Access Bearer (RABs).

5.8.11 Radio Resources Control (RRC) Server States

Because of the interactively of the PoC service, it is important to keep the response time short. Based on the user activity the terminal is in different Radio Resource Control (RRC) states: Cell DCH, Cell FACH, Cell PCH or URA PCH. In order to keep the response times short also when the terminal has been inactive for some time between PoC communication sessions, it is important to support and is not only the Cell DCH and Cell FACH states, but also at least the Cell PCH state.

5.8.12 WCDMA Capacity Enhancements and PoC

The Degermark Header Compression is similar in WCDMA and (E) GPRS in terms of decreasing the throughput and saving the radio interface capacity. However, in WCDMA the compression takes place in the Radio Network Controller (RNC), while in E(GPRS) it is the 2G SGSN that handles the compression.

The Robust Header Compression is an IP/UDP/RTP header compression protocol that is suitable for wireless links as it guarantees high compression efficiency and performs well over links with high error rates.

5.8.13 HSDPA and PoC

High Speed Downlink Packet Access (HSDPA) is part of 3GPP Release 5. It provides several improvements, for example, for the average end user bit rates, aggregate cell throughputs, and packet round trip times. From an end user point of view, the main improvement with HSDPA is expected to be a slight reduction of the response time. From an operator point of view, the HSDPA shared channel approaches may enhance the control of the resources and hence enable the introduction of even larger PoC group sizes, particularly in highly loaded networks.

5.9 Summary and Conclusions

In this chapter we have presented a set of technical and commercial issues related to deployment of PoC services in a real environment. Particular focus has been put on the PoC over IMS architecture, and numerous concepts have been clarified. We have seen how different PoC and IMS concepts are related, and how it is feasible to deploy PoC over IMS in a consistent and efficient way. Although technical challenges will arise in any deployment, it is clear that the modular architectures defined by 3GPP and OMA can interoperate through standard interfaces, and deliver an end-to-end service to the final user in a smooth way.

The following itemized list provides a non-exhaustive checklist of topics covered in this chapter, which may be useful for service providers and operators to ensure that all elements in the value chain are in place to enable PoC service delivery and satisfactory user experience:

- Radio network parameters need to accommodate IP based group communication;
- A mechanism to provision PoC handset information must be in place;
- The IMS must provision a Public User Identity for those users accessing the PoC service from a device equipped with SIM/USIM cards (i.e. without ISIM card);
- A convention should be established to align PoC identity with XUI, so that documents stored in the XCAP tree are consistent with the identity used for the PoC service;
- A naming convention for PoC groups should be defined, aligned with PSI's and provisioned in the XDMC as the URI template;
- Originating and Terminating iFC should be configured for PoC, Presence and XDM;
- A billing model should be defined. This should be mapped to the technical charging triggers required for online charging;
- The S-CSCF should be configured to provide the relevant addresses of charging nodes (e.g. OCS, CDF) to application servers responsible for triggering charging operations (e.g. the PoC AS);
- The service provider should select the format of generate CDRs and ensure that enough information is provided by the PoC AS through Rf interface;
- Third Party Registration or Subscription to the Registration Event package should be configured if needed;
- The ENUM service should be configured, and the operator should decide where ENUM queries take place (i.e. S-CSCF or PoC AS).
- A consistent 2.5G (e.g. EGPRS), 3G (WCDMA) or 3.5G (e.g. HSDPA/HSUPA) radio resource management and planning strategy should be put into place to enable an acceptable and homogeneous service level to PoC end-users, as a real-time VoIP service over the cellular infrastructure.

5.10 References

[1] M. Poikselkaa, A. Niemi, H. Khartabil, G. Mayer: 'The IMS: IP Multimedia Concepts and Services', John Wiley & Sons, March 2006.
[2] G. Camarillo, M. A. García-Martín: 'The 3G IP Multimedia Subsystem'; John Wiley & Sons, February 2006.
[3] Open Mobile Alliance: 'Instant Messaging using SIMPLE Architecture', Draft Version 1.0; November 2006 (work in progress).

[4] 3GPP TS 23.002 v7.1.0: 'Network Architecture (Release 7)', March 2006.

[5] H. Kaaranen, A. Ahtiainen, L. Laitinen, et al.: 'UMTS Networks: Architecture, Mobility and Services', John Wiley & Sons, April 2005 (2ⁿᵈ Edition).

[6] H. Holma, A. Toskala: 'WCDMA for UMTS: Radio Access for Third Generation Mobile Communications', John Wiley & Sons, September 2004 (3ʳᵈ Edition).

[7] H. Schulzrinne: 'The Tel URL for Telephone Numbers'; RFC 3966; December 2004.

[8] J. Rosenberg, H. Schulzrinne, G. Camarillo, et al.: 'SIP: Session Initiation Protocol'; RFC 3261; June 2002.

[9] OMA Push-To-Talk over Cellular (PoC v1.0.1): 'PoC Control Plane'; November 2006.

[10] 3GPP 23.003 v7.2.0: 'Numbering, addressing and identification (Release 7)', December 2006.

[11] OMA XDMv1.0.1: 'XDM Core Technical Specification', November 2006.

[12] P. Faltstrom, M. Mealling: 'The E.164 to Uniform Resource Identifiers (URI) Dynamic Delegation Discovery System (DDDS) Application (ENUM)', RFC 3761, April 2004.

[13] OMA Push-To-Talk over Cellular (PoCv1.0.1): 'PoC Architecture'; November 2006.

[14] 3GPP TS 23.218 v6.4.0: 'IP Multimedia (IM) session handling; IM call model; Stage 2 (Release-6)', June 2006.

[15] 3GPP TS 29.228 v6.13.0: 'IP Multimedia (IM) Subsystem Cx and Dx Interfaces; Signalling flows and message contents (Release-6)', December 2006.

[16] 3GPP TS 23.221 v6.3.0: 'Architectural Requirements (Release-6)', June 2004.

[17] OMA Push-To-Talk over Cellular (PoC v1.0.1): 'PoC User Plane'; November 2006.

[18] 3GPP TS 32.200 v5.9.0: 'Telecommunication management; Charging management; Charging principles (Release-5)', September 2005.

[19] 3GPP TS 32.240 v6.4.0: 'Telecommunication management; Charging management; Charging architecture and principles (Release-6)', September 2006.

[20] 3GPP TS 32.260 v6.8.0: 'Telecommunication management; Charging Management; IP Multimedia Subsystem (IMS) charging (Release-6)', March 2007.

[21] 3GPP TS 32.272 v6.5.0: 'Telecommunication management; Charging management; Push-to-talk over Cellular (PoC) charging (Release-6)', October 2006.

[22] P. Calhoun, J. Loughney, E. Guttman, et al.: 'Diameter Base Protocol'; RFC 3588, September 2003.

[23] 3GPP TS 32.295 v6.1.0: 'Telecommunication management; Charging management; Charging Data Record (CDR) transfer (Release-6)', June 2005.

[24] H. Hakala, L. Mattila, M. Stura, et al.: 'Diameter Credit-Control Application'; RFC 4006, August 2005.

[25] 3GPP 32.296 v6.3.0: 'Telecommunication management; Charging management; Online Charging System (OCS): Applications and interfaces (Release-6)', October 2006.

[26] M. Gacía Martín, E. Henriksson, D. Mills: 'Private Header (P-Header) Extensions to the Session Initiation Protocol (SIP) for the 3rd-Generation Partnership Project (3GPP)'; RFC 3455, January 2003.

[27] J. Rosenberg: 'A Session Initiation Protocol (SIP) Event Package for Registrations'; RFC 3680, March 2004.

[28] OMA Presence (Presence v1.0.1): 'Presence SIMPLE Technical Specification'; November 2006.

[29] OMA Device Management (Device Management v1.2), February 2007.

[30] 3GPP TR 23.979 v6.2.0: '3GPP enablers for Open Mobile Alliance (OMA) Push-to-Talk over Cellular (PoC) services; Stage 2 (Release-6)', June 2005.

[31] J. Sjoberg, M. Westerlund, A. Lakaniemi, et al.: 'Real-Time Transport Protocol (RTP) Payload Format and File Storage Format for the Adaptive Multi-Rate (AMR) and Adaptive Multi-Rate Wideband (AMR-WB) Audio Codecs'; RFC 3267, June 2002.

[32] C. Bormann, C. Burmeister, M. Degermark, et al.: 'RObust Header Compression (ROHC): Framework and four profiles: RTP, UDP, ESP, and uncompressed'; RFC 3095, July 2001.

[33] 3GPP TS 23.203 v7.2.0: 'Policy and charging control architecture (Release-7)', March 2007.

[34] 3GPP 33.978 v6.5.0: 'Security aspects of early IP Multimedia Subsystem (IMS) (Release 6)', September 2006.

6

Examples of Group Communication Sessions

6.1 Introduction

This chapter gives examples of real-time group communication and applies the concepts of OMA PoC and OMA list and policy management. The examples focus on a PoC user called John, who engages in different types of group sessions. John's group communication is facilitated by two service providers who have set up a Service Level Agreement. John is a subscriber to the service provider 'Wirelessfuture.com'. Most of John's contacts are subscribers of a competitive service provider called 'OtherDomain.com'. OtherDomain.com has built up a community of PoC users, who utilise group communication to achieve their individual aims.

The examples are:

- John registers to Wirelessfuture.com's IMS/PoC service.
- John creates an Ad-hoc session with Mary during a robotics conference. John has just met Mary and wants to meet her again in the evening. He uses PoC, since it is cheaper than making two or more GSM calls. Next, John decides to invite Mary's friend Alice to the call, and adds her to the Ad-hoc group session.
- Paul creates a Pre-arranged group session with fellow team members of Chatham Wanderers on Friday, the day before an important qualifying match. Alf coaches Chatham Wanderers. Alf uses PoC as a cheap and efficient communication tool to outline the players' positions and roles and to give an update on the opposition's form. Paul is the captain of Chatham Wanderers and has the rights to start up a Pre-arranged session when Alf is busy.
- John is a part time taxi driver and joins an open taxi Chat group when he starts his shift with Fast Taxi. All the drivers of Fast Taxi use PoC instead of expensive and bulky walkie-talkies to organize collecting their regular customers.
- Alice is a dispatcher for a VIP casino. The casino uses PoC as a cost effective tool to organize collecting important guests, for welcoming guests, controlling the VIP room,

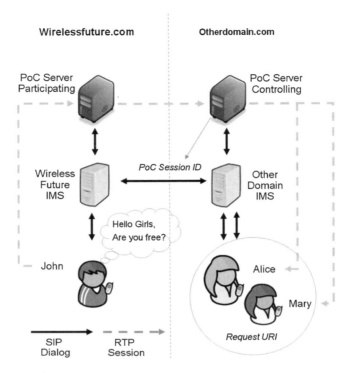

Figure 6.1 Signalling flow and media exchange paths for Network to Network scenario

and ensuring the smooth running of the casino. Alice sends a Group Advertisement to Mary.

This chapter presents signalling flows and basic procedures for SIP and TBCP adopted from OMA PoC architecture, control plane and user plane specifications (V1.0.1) [1–3]. For the SIP traffic, the reader should note that the Request URI identifies the PoC service, PoC group identities or PoC participants, and PoC Session ID is used to route SIP requests. Exchange of media occurs within an RTP session, and CODECs and session description parameters have to be successfully negotiated end-to-end to ensure full interoperability between different vendors' mobile devices and servers. The next subsection will define these important attributes before moving on to the examples. The relation between the signalling flow and media path between the two networks are shown in Figure 6.1.

6.1.1 Signalling and media paths

- A SIP session is a SIP dialog. From RFC 3261, a SIP dialog is defined as 'a peer-to-peer SIP relationship between two User Agents that persists for some time. A dialog is established by SIP messages, such as an OK response to an INVITE request. A dialog is identified by a SIP Call-ID, local tag, and a remote tag'. All related SIP messages use the same Call ID, for example SIP INVITE until SIP BYE. The SIP CSeq header is used to keep track of the sequence of messages exchanged within Call-ID.

6.1.2 Request-URI

The Request URI is issued by the originating PoC client and identifies the target destination for an Ad-hoc conference call or for a group call. The Request URI can be:

- The PoC conference factory address when used for Ad-hoc sessions. The PoC conference factory address uniquely identifies the PoC server, for example: 'sip:Ad-hoc@Wirelessfuture.com';
- The PoC group identity when session type is 'Pre-arranged';
- The PoC group identity when session type is 'Chat';
- The PoC group identity or a PoC user identity when sending a Group Advertisement;
- The PoC user identity when sending an Instant Personal Alert (IPA).

When adding a user or rejoining an ongoing session, the Request URI is set to the existing PoC session identity.

6.1.3 PoC Session Identity

A PoC Session Identity SIP URI identifies the PoC session and is used for routing initial SIP requests. The PoC client receives the PoC Session ID during the PoC session establishment in the Contact header and/or in the TBCP Connect message in case of using Pre-established session. The PoC server performing Controlling PoC function allocates a unique PoC Session Identity for the PoC session when established. The PoC server performing the Participating PoC function can modify the PoC Session Identity.

The PoC Session Identity identifies the PoC session to the extent that:

- The PoC user is able to add PoC users to an ongoing PoC session;
- The PoC user is able to rejoin the PoC session, as long as the PoC session is ongoing in the PoC server performing the Controlling PoC function;
- The PoC user is able to subscribe the Participant information of the ongoing PoC session;
- The PoC user is able to leave a PoC session;
- The SIP/IP Core enables routing of initial SIP requests to the PoC server performing the Controlling PoC function.

The PoC server performing Controlling PoC function sends the PoC Session Identity towards the PoC client during the PoC session establishment in the Contact header.

The PoC server performing Participating PoC function sends the PoC Session Identity to the PoC client in the TBCP Connect message if a Pre-established session is used.

The PoC Session Identity contains a globally unique identifier for a call, generated by the concatenation of a random string and the hostname or IP address [4]. An example Contact header containing a Session Identity could be:

Contact: <sip:d4it34fvh59diov0v8@Wirelessfuture.com;session=Pre-arranged>.

6.1.4 RTP Session

An RTP session is an association that allows exchange of RTP media streams and RTCP messages among a set of PoC functional entities. The RTP sessions are kept unique by the packet sequence number, timestamp and synchronization source (SSRC). The SSRC is a randomly chosen value and no synchronization source within the same RTP session will have the same SSRC. If a source changes source address, it must also choose a new SSRC. In this chapter, the RTP sessions are established between:

- Originating PoC client (John) and Wirelessfuture PoC server
- Wirelessfuture PoC server and OtherDomain PoC server
- OtherDomain PoC server and OtherDomain PoC clients.

6.1.5 CODECs

3GPP mandates the AMR narrowband speech CODEC as the default speech CODEC for PoC service. However, 3GPP also mandates support of the AMR wideband speech CODEC, if the User Equipment on which the PoC client is implemented uses 16 kHz sampling frequency of the speech. The media parameters for the AMR narrowband speech CODEC and the AMR wideband speech CODEC are described together with the RTP payload format for the speech CODECs in [RFC3267] = [5].

OMA PoC compliant clients must support all eight AMR CODEC modes, which operate between 4.75 kbps and 12.2 kbps (CODEC rate, excluding RTP, UDP, IP and Layer-2 header overhead). However, most commercial deployments use only the 4.75 and 5.15 kbps modes. This is in order to enable PoC over 2.5G, when the available bandwidth may be as low as 8 kbps (e.g. GPRS network allocating one timeslot for the PoC service).

3GPP2 mandates the EVRC speech CODEC when PoC is deployed in 3GPP2-based networks (e.g. cdma2000). Encapsulation of EVRC CODEC over RTP is described in [6]. EVRC may operate at two different rates, namely: 4.0 kbps and 8.55 kbps, with a third comfort noise generation mode at 0.8 kbps.

6.1.6 RTP and TBCP Session description parameters

The OMA PoC client that supports 3GPP AMR speech CODEC constructs an SDP answer for a payload type in an SDP offer with the following parameters [7]:

- octet-align=1 or octet-align=0 (default value) parameter. This parameter indicates whether RTP fields are octet-aligned in the RTP AMR header (easier parseability), or not (more efficient encoding).
- ptime and maxptime parameters are used to indicate the typical (ptime) or maximum (maxptime) time duration of an RTP packet carrying AMR payload. OMA PoC clients should support a maxptime value of up to 400 ms.
- crc=0 or no crc parameter. This parameter can be used to detect bit errors in an RTP packet. Since PoC is a packet based service, this parameter is set to 0 or omitted from the session description.
- The robust sorting parameter serves a purpose similar to crc, when *Unequal Error Protection* (UEP) is applied to AMR data. This parameter is not used in the PoC service.
- The interleaving parameter is not used. The goal of this setting is to enhance reliability by randomizing error bursts. To achieve this, audio frames are interleaved, thus incurring an

extra delay in the end-to-end transmission. Hence, frame interleaving is disabled in order to avoid excessive end-to-end delay for a VoIP service such as PoC, which relies on tight delay requirements, as described in [8].

• The number of channels is 1, since AMR is a single channel CODEC.

Additional mandatory SDP parameters must be used to describe the PoC session, and other optional parameters may be used as well, as described in [9]. Consequently, the following paragraph shows a sample SDP offer that could be constructed by an OMA PoC compliant client initiating a group call.

```
v=0
o=John 2890844526 2890844526 IN IP4 WirelessFuture.com
s=PoC Session
c=IN IP6 5555::aaa:bbb:ccc:ddd
t=0 0
m=audio 3456 RTP/AVP 96
a=rtpmap:96 AMR/8000
a=fmtp=97 mode-set=0,1,2,3
a=rtcp:3457
m=application 3457 udp TBCP
a=fmtp:TBCP queuing=1;tb_priority=2; timestamp=1
```

In the above session description, the UE signals its capability to support AMR modes 0, 1, 2 and 3 (i.e. audio encoded at 4.75, 5.15, 5.9 and 6.7 kbps and it indicates the desired local UDP port for such traffic (3456). The source and destination address for media traffic is retrieved in the (c=) connection header, which contains an IPv6 address.

In addition to the RTP flow, John's OMA PoC device must indicate its capability to support the *floor control* protocol (TBCP), which – from the SIP and SDP signalling point of view – is seen simply as an additional media type. For this, the client inserts a new media description, indicating that the TBCP flow will use port 3457. Since TBCP is based on RTCP [10], generally the RTCP port convention is used (the UDP port for RTCP traffic is generally one unit greater than the UDP port used for RTP traffic).

John's PoC device can also signal its support and willingness to implement some optional OMA features, such as floor control message queuing, priority support and the ability to include optional timestamp information in TBCP Talk Burst Request messages. Timestamps enable for finer granularity when deciding which user has the right to speak (in case of collision of floor requests from two or more session participants).

6.2 PoC Service Registration

When John registers to the SIP/IP core, his PoC client generates a SIP REGISTER request that includes: PoC feature tags '+g.poc.talkburst' and '+g.poc.groupad' in the Contact header. These indicate to the SIP/IP Core that – possibly among other service – this UE is capable of supporting the PoC service. The SIP REGISTER message also includes a Require header with the option tag 'pref' and a User-Agent header to indicate the PoC client release and version. The Require header is used to mandate that the server processing this request will understand and support user preferences, as indicated in subsequent SIP messages sent by the UE.

6.2.1 PoC Settings

To set John's PoC Service Settings, the PoC client generates a SIP PUBLISH and sets the Request-URI to the PoC address of the PoC user. The SIP PUBLISH includes the PoC address of the PoC user as the Authenticated Originator's PoC address. An Accept-Contact header with the PoC feature tag '+g.poc.talkburst' along with 'require' and 'explicit' parameters is inserted, to indicate to the SIP network that this client is willing to be contacted via PoC means. The sigalling flow for PoC registration is shown in Figure 6.2.

The basic procedures of registration to a PoC services are:

1. **SIP REGISTER request from John's PoC client to SIP/IP Core**. John's PoC client sends SIP REGISTER request containing the PoC feature tag '+g.poc.talkburst' in the contact header and the authentication response in the Authorization header to Wirelessfuture's SIP/IP Core.

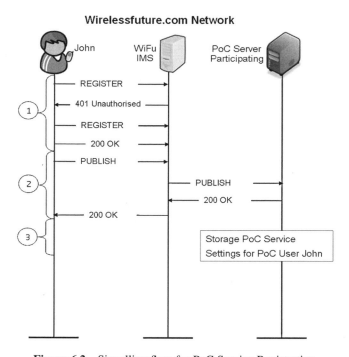

Figure 6.2   Signalling flow for PoC Service Registration

```
REGISTER sip:poc.wirelessfuture.com SIP/2.0
From: <sip:John@poc.wirelessfuture.com>;tag=4fa3
To: <sip:John@poc.wirlessfuture.com>
User-Agent:PoC-client/OMA1.0
Contact: <sip:John@poc.wirelessfuture.com>;
+g.poc.talkburst; +g.poc.groupad
```

The IMS sends SIP 401 'Unauthorized' responds to challenge and authenticate John's PoC client. The client issues a new SIP REGISTER request with authentication credentials and, eventually, the PoC client gets registered into the SIP/IP Core network.

Observe that, at this stage, some SIP/IP Core implementations such as 3GPP IMS may forward the successful REGISTER request towards trusted Application Servers (such as the PoC server) to inform them about the registration status of local users. This is not depicted in Figure 6.2.

2. **SIP PUBLISH request from John's PoC client to SIP/IP Core.** John's PoC client publishes the current PoC Service Settings of PoC client by sending a SIP PUBLISH request for the event package 'poc-settings' to SIP/IP Core. The Request-URI is sip:John@poc.wirelessfuture.com

```
PUBLISH sip:John@poc.wirelssfuture.com SIP/2.0
P-Preferred-Identity: 'John' <sip:John@poc.
  wirelessfuture.com>;
Accept-Contact: *;+g.poc.talkburst; require;explicit
User-Agent:PoC-client/OMA1.0
Contact: <sip:John@poc.wirelessfuture.com>;
Event: poc-settings
Content-Type: application/poc-settings+xml
```

The IMS forwards the SIP PUBLISH to PoC server. The PoC server responds with SIP 200 'OK' to the IMS. The IMS sends SIP 200 'OK' response to John's PoC client.

1. **PoC server stores the PoC Service Settings** for John's PoC client

The SIP PUBLISH message sets the Event header to the value 'poc-settings' and includes the PoC Service Settings. This event consists of the publication of an XML body that indicates to the PoC server the PoC settings to apply to the publishing user.

The four parameters that can be configured using the poc-settings event are:

- Incoming Session Barring flag. When enabled, incoming session invitations from other PoC users are automatically rejected at the PoC server.
- Answer Mode Setting. This parameter regulates whether incoming sessions will be automatically accepted, or user manual acceptance is required first.

- Incoming Alert Barring flag. This parameter determines whether incoming IPA (Call-Back Request) messages are accepted or directly rejected by the PoC server.
- Simultaneous Session Support flag. This setting indicates whether the client is capable and willing to support participation in several PoC sessions in parallel.

In the following example, John has set Incoming Session Barring (ISB) as active; set the answer mode to Automatic; set Instant Personal Alert Barring as active; and his client supports Simultaneous sessions.

```
<?xml version='1.0' encoding='UTF-8'?>
<poc-settings xmlns='urn:oma:params:xml:ns:poc:poc-settings'
 xmlns:xsi='http://www.w3.org/2001/XMLSchema-instance'
 xsi:schemaLocation='urn:oma:params:xml:ns:poc:poc-settings'>
 <entity id='do39s8zksn2d98x'>
  <isb-settings>
    <incoming-session-barring active='true'>
  </isb-settings>
  <am-settings>
    <answer-mode>automatic</answer-mode>
  </am-settings>
  <ipab-settings>
    <incoming-personal-alert-barring active='true'/>
  </ipab-settings>
  <sss-settings>
    <simultaneous-sessions-support active='true'/>
  </sss-settings>
 </entity>
</poc-settings>
```

Observe that, since the ISB flag is enabled, the answer mode setting does not have any effect on incoming PoC session invitations.

John is now online with Wirelessfuture's PoC service.

6.3 Ad-hoc Group Session

During a robotics conference, John decides to initiate an On Demand Ad-hoc PoC group session with Mary. Mary is a subscriber to the OtherDomain.com network (as shown in Figure 6.3).

Acting on his behalf, John's PoC client generates an initial SIP INVITE and sets the Request-URI request to the Conference-factory-URI for the Wirelessfuture's PoC service. The SIP INVITE message incorporates a multipart body, comprised of an SDP content and a resource list XML content.

As described above, the SDP message is used to:

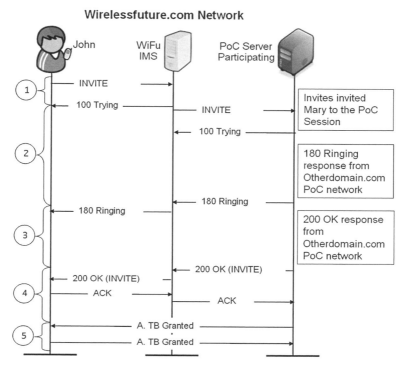

Figure 6.3   Signalling flow for starting an Ad-hoc session

- Set the IP address and port number for the RTP session;
- Include the CODEC(s) and media parameters being offered by the PoC client for the PoC service;
- Set IP address and port number to be used for RTCP at the PoC client;
- Talk Burst parameter(s) and the port number(s) for the Talk Burst Control Protocol(s).

The resource list provides the list of PoC users (identified by their corresponding PoC identity, either a SIP URI or a TEL URL) to be invited to participate in the Ad-hoc session.

In our example, initially, one single PoC user (Mary) is invited to participate in the Ad-hoc session, which will effectively be a one-to-one PoC call.

The SIP INVITE is sent to the Wirelessfuture SIP/IP core (1); forwarded to the Wirelessfuture's PoC service (2); sent to OtherDomain.com SIP/IP Core (3); forwarded to Other-Domain.com PoC service; and finally sent Mary's PoC client, which will eventually accept the session invitation (4).

The basic procedures of setting up an Ad-hoc group session are:

1. **The PoC client instructions.** John's PoC client sends an SIP INVITE to Wirelessfuture's IMS (SIP/IP core). The Request-URI is sip:Ad-hoc@.poc.wirelessfuture.com. The PoC client SIP INVITE message contains:

```
INVITE sip:PoCConferenceFactoryURI@Wirelessfuture.com
  SIP/2.0
P-Preferred-Identity:  'John'  <sip:John@poc.
  wirelessfuture.com>
Accept-Contact: *;+g.poc.talkburst; require;explicit
User-Agent:PoC-client/OMA1.0
Privacy: id
Contact:   <sip:John@poc.wirelessfuture.com>;+g.poc.
  talkburst
Supported: Timer
Session-Expires: 1800;refresher=uac
Allow: INVITE,ACK,CANCEL,BYE,REFER,
Content-Type: multipart/mixed
```

The inserted SDP content will set up the media stream between the entities:

```
v=0
o=John 2890844526 2890844526 IN IP4 WirelessFuture.com
s=PoC Session
c=IN IP4 10.10.28.42
t=0 0
m=audio 65304 RTP/AVP 106
a=rtpmap:106 AMR/8000
a=fmtp=106 mode-set=0,1,2,3; octet-align=1
a=maxptime:400
a=ptime:160
a=rtcp:3457
m=application 6305 udp TBCP
a=fmtp:TBCP queuing=0;tb_priority=1; timestamp=0
```

The above session description provides basic connectivity, CODEC and talk burst flow setup. When compared to the example shown in section 7.1.6, we observe that:

- The PoC client indicates its preference to receive audio data in chunks of 160 ms of duration, but it is capable of accepting up to 400 ms of encoded audio in a single packet.
- The client does not support TBCP message queuing.

The inserted MIME resource lists body with the PoC address of the Invited PoC user (Mary), as follows:

```
Content-Type: application/resource-lists+xml
Content-Disposition: recipient-list
<?xml version='1.0' encoding='UTF-8'?>
<resource-lists xmlns='urn:ietf:params:xml:ns:
```

```
resource-lists'
xmlns:xsi='http://www.w3.org/2001/XMLSchema-instance'>
 <list>
   <entry uri='sip:Mary@poc.otherdomain.com' />
 </list>
</resource-lists>
```

The Wirelessfuture IMS replies with SIP 100 'Trying' response to John's PoC client.

2. **The PoC service contacts the group participants.** The IMS sends a SIP INVITE to Wirelessfuture PoC server. The Preferred Identity is modified by the IMS to an Asserted Identity, so that IMS nodes can trust the identity of the authenticated user:

```
P-Asserted-Identity: 'John' <sip:John@poc.wirelessfuture.
com>...
```

The Wirelessfuture PoC server responds to the IMS core with a SIP 100 Trying, to stop client SIP INVITE retransmission timers. The Wirelessfuture PoC server sends SIP INVITE requests towards Mary's client, the Request-URI, P-Asserted Identity and its media parameters:

```
INVITE   sip:Mary@ poc.otherdomain.com SIP/2.0
P-Asserted-Identity:  'John'  <sip:John@poc.wirelessfuture.
  com>
Referred-By:  'John'  <sip:John@poc.wirelessfuture.com>
```

3. **The PoC service waits for a response from the group members.** The PoC server receives SIP 180 'Ringing' response:

```
SIP/2.0 180 RingingServer: PoC-serv/OMA1.0
Contact: <sip:PoC-Session12345@com.poc.server.otherdomain.
  com;session=1-1>;+g.talkburst
```

The response contains a Contact: header that identifies the PoC session at Otherdomain. com. The Wirelessfuture PoC server forwards the SIP 180 'Ringing' response to the IMS. The response contains a new Contact header that provides the PoC session identity to John's PoC client.

```
SIP/2.0 180 Ringing
Server: PoC-serv/OMA1.0
Contact: <sip:PoC-SessionABCD@poc.server.wirelessfuture.
  com;session=1-1>;+g.talkburst;isfocus
```

The 'isfocus' parameter indicates that the PoC session identity carried in the Contact header has the conference focus.

The IMS forwards the SIP 180 'Ringing' message to Wirelessfuture PoC client server:

4. **Group Ad-hoc session is verified.** The PoC server receives SIP 200 'OK' response from OtherDomain.com, and sends the SIP 200 OK to IMS. The 200 OK message carries the allocated PoC session identity in the Contact header.

```
SIP/2.0 200 OK
Server: PoC-serv/OMA1.0
Contact: <sip:PoC-SessionABCD@poc.server.wirelessfuture.
  com;session=1-1>;+g.talkburst;isfocus
```

The IMS sends a SIP 200 'OK' to John's PoC client. John's PoC client returns a SIP ACK to the IMS. The IMS forwards it to the PoC server.

5. **Media exchange between group participants**. The PoC server sends a TBCP Talk Burst Granted message to John (alternatively, the Talk Burst Granted indication can be provided within the SDP body carried in the 200 OK message delivered to John's client). John's PoC client sends RTP media to PoC server, which is sent up to Mary's PoC client.

6.3.1 Add a participant to an Ad-hoc group session

John decides to add Alice to the ongoing PoC session with Mary. John selects Alice from his contact list and adds her to the session. To add Alice to the session, John's PoC client generates an initial SIP REFER and sets the Request-URI to the PoC Session Identity of the current PoC session. Once Alice accepts to participate in the session, the OtherDomain.com Controlling PoC server issues a SIP NOTIFY and the communication path is set up. The signalling flow is shown in Figure 6.4.

The basic procedures of adding an additional member to the Ad-hoc group session are:

1. John's PoC client sends a SIP REFER request with the intended invited PoC user's address (Alice) through the signalling path to the OtherDomain.com PoC server (Controlling function). The SIP REFER request contains the PoC address of the invited PoC user (Alice) and the PoC Session Identity (sip:PoC-SessionABCD@poc.server.wirelessfuture. com;session=1-1) of the PoC session to add.

2. Upon receiving the SIP REFER request, OtherDomain.com PoC server (Controlling function) initiates the inviting procedure to Alice's PoC client and sends a '202 Accepted' response to John's PoC client through the signalling path which routed the original request.

When OtherDomain.com's PoC server (Controlling) receives the indication from Alice's PoC client that it has accepted the invitation, OtherDomain.com PoC server (Controlling) sends a SIP NOTIFY request to John's PoC client, to indicate that Alice has just joined the call. John's PoC client sends an OK response to the OtherDomain.com PoC server (Controlling).

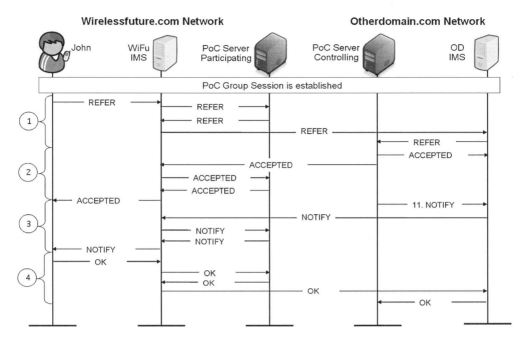

Figure 6.4 Signalling flow for adding a user to an ongoing Ad-hoc session

6.3.2 Session Termination

Following the discussion with Mary and Alice, John terminates the session when he leaves. John's PoC client generates a SIP BYE and upon receipt, the PoC server releases the PoC session. The general signalling flow for leaving a PoC session is shown in Figure 6.5.

The basic procedures of terminating an Ad-hoc group session for Figure 7.5 are:

1. John's PoC client sends a SIP BYE request to the Wirelessfuture IMS. The Request-URI contains the session identity – sip:PoC-Session 12345@pocserver.OtherDomain.com@ Wirelessfuture.pocserver.wirelessfuture.com. The Wirelessfuture IMS sends the SIP BYE request to the Wirelessfuture PoC server.
2. The Wirelessfuture PoC server sends a SIP 200 'OK' to the Wirelessfuture IMS. The IMS forwards the SIP 200 'OK' message to the PoC client.
3. Wirelessfuture IMS sends SIP BYE request to OtherDomain IMS. The request URI contains PoC-Session-Identity12345@com.pocserver.otherdomain.com. The OtherDomain IMS sends SIP BYE to the OtherDomain PoC server.
4. OtherDomain PoC server sends a SIP 'OK' to the OtherDomain.com IMS. Eventually, Wirelessfuture PoC server receives a 200 'OK' message from OtherDomain.com Controlling PoC server through OtherDomain.com and Wirelessfuture.com IMSs. The PoC session is terminated.

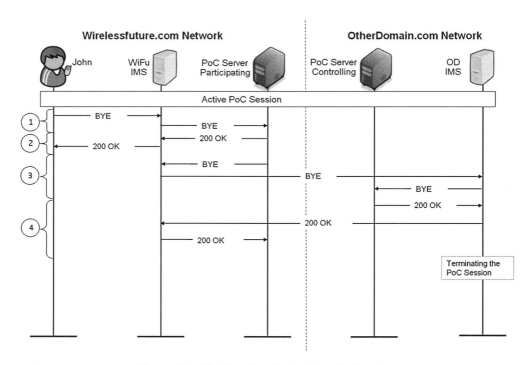

Figure 6.5 Sigalling flow for leaving a PoC session

6.4 Pre-arranged Group Session

Paul is the captain team of a football team called Chatham Wanderers. Chatham Wanderers is coached by Alf, who uses PoC to coordinate training sessions and to broadcast information updates. Alf has created a Pre-arranged group called 'Chatham Wanderers' and he has given Paul the rights to initiate the group session when he is unavailable. In the following example, Alf's PoC document is briefly described. John initiates a Pre-arranged group session, drops out and rejoins a session. The Chatham Wanderers Pre-arranged group is owned by Other-Domain PoC server (Controlling function).

PoC groups are defined in XML documents whose format and semantics is specified in [11]. In this section we provide an example explaining how the different XML elements are combined to define the actions that the PoC server should take when a PoC group session is initiated.

6.4.1 XML PoC Group Document

Alf is the owner of an XML PoC document called 'Chatham_Wanderers.xml' and he has given group members Paul and John management communication rights to act on his behalf (Figure 7.6). Such rights are part of the document meta-data and defined as follows:

- **Creator** – Alf is the original creator;
- **Primary principal** – Alf is the original creator of document with permissions and is able to retrieve and manage the Chatham Wanderers document from the OtherDomain.com PoC service;

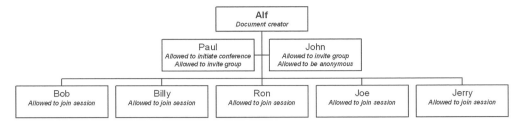

Figure 6.6   Alf's Pre-arranged Chatham Wanderers team

- **Permissions** – Alf has given permission rights to perform operations on his behalf to the other principals;
    - Paul is allowed to initiate a Chatham Wanderers conference sessions and invite team members who have not accepted to the initial session set up;
    - John is allowed to invite team members and is allowed to be anonymous during the Pre-arranged group session;
- **Document Identifier** is Alf.Garnet.Chatham_Wanderers.xml;
- **Visibility** – Principals can find Alf's document when performing a search.

The policies outlined in Alf's Chatham Wanderers document instruct the OtherDomain.com PoC Controlling function how to set up the Chatham Wanderers Pre-arranged group session. Three important policies define the Chatham Wanderers Pre-arranged group sessions.

- **Allow-initiate-conference policy** – Within Alf's Pre-arranged Chatham Wanderers PoC Group document, Paul's PoC Addresses is allowed the action **<allow-initiate-conference>**. When Paul initiates the Pre-arranged group session, the Controlling PoC function invites PoC group members contained in the **<list>** element of Alf's PoC Group Document. If the number of PoC group members exceeds **<max-participant-count>**, the Controlling PoC function invites only <max-participant-count> members from the list.
- **Allow-invite-users-dynamically policy** – Within Alf's Pre-arranged Chatham Wanderers PoC Group Document, John's and Paul's PoC addresses are allowed by the **<allow-invite-users-dynamically>.** When John or Paul request to add one or more PoC users to an ongoing Chatham Wanderers PoC session, the Controlling PoC function only invites PoC users if the Invited team players are listed in the **<list>** element of the PoC Group Document.
- **Joining-handling policy** – The PoC server performing the Controlling PoC function will allow only those PoC users to join in the Pre-arranged or Chat PoC session that it hosts, whose Authenticated Originator's PoC address is allowed by the **<join-handling>** action of the PoC group's authorization rules. In Alf's Pre-arranged Chatham Wanderers PoC document, all team members are allowed to join an ongoing session.

A complete description of the XML elements that set the PoC Controlling functions is listed in Table 6.1.

6.4.2 Pre-arranged PoC Group Session Initiation

Paul initiates a Pre-arranged PoC group session for Chatham Wanderers. When he does this, all other team members are invited. The Pre-arranged PoC group session is established

Table 6.1   XML elements that define group communication

XML Element	Explanation
⟨list-service⟩	Which contains the SIP URI attribute representing the PoC Group Identity

The ⟨list-service⟩ element may include:

⟨display-name⟩	Display a human readable name of the group '**Chatham Wanderers**'.
⟨list⟩ element	Contains the PoC Group Members -- contains an attribute URI which contains a valid PoC address (SIP URI or TEL URI).
⟨invite-members⟩	Indicates whether the PoC server will invite the PoC group member to the PoC group session. A true represents the Pre-arranged group. The PoC server performing the Controlling PoC function will invite the member in the ⟨**list**⟩ element.
⟨max-participant count⟩	Indicates the maximum number of Participants allowed by the document owner in the PoC group session.

The ⟨list-service⟩ includes a ⟨ruleset⟩ the authorization policy associated with this PoC group document. The ⟨ruleset ⟩ include ⟨conditions⟩ child elements of any ⟨rule⟩:

⟨external list⟩	References URI Lists stored in the Shared XDMS. Such reference URI list belongs to the same user as that of the PoC group document.
⟨is-list-member⟩	Is used to match an identity against the content of the ⟨list⟩ element.

The ⟨actions⟩ child element of any ⟨rule⟩ elements may include child elements:

⟨allow-conference-state⟩	Indicates whether the identity matching this rule is allowed to subscribe to the 'conference' event package or not. The 'conference' event packages provide real-time information about status of session participants.
⟨allow-invite-users-dynamically⟩	Indicates to the PoC server performing the Controlling PoC function that inviting additional participants is allowed. A true value instructs the PoC server to allow the user (Alice) to invite additional participants.
⟨join-handling⟩	Defines the action that the PoC server performing the Controlling function is to take when processing a request to join a PoC group session. A true value instructs the PoC server to accept access to the PoC session.
⟨allow-initiate-conference⟩	Allows users (**Paul**) to initiate a Pre-arranged PoC group session. A true value instructs the PoC server to allow the user to initiate the Pre-arranged PoC group session.
⟨allow-anonymity⟩	Indicates whether anonymity is allowed for a matching identity that is requesting anonymity (**John**). When true, it instructs the PoC server to accept an anonymous access to the PoC group session
⟨is-key-participant⟩	Indicates that the identity matching this rule is a 'Distinguished Participant'. It instructs the PoC server to treat the user as a Distinguished Participant if the one-to-many-to-one topology is used.

by using the group identity in the invitation message. The signalling flow is shown in Figure 6.7.

The basic procedures of setting up a Pre-arranged group session are:

1. **Setting up the Pre-arranged session**. Paul's PoC client sends Wirelessfuture IMS a SIP INVITE request to the address of the Pre-arranged PoC group (sip:Chathamwanderers@ OtherDomain.com). The SIP INVITE contains the Pre-arranged PoC group identity, PoC address of Paul's PoC client, PoC service indication, media parameters of Paul's PoC client, and Talk Burst Control Protocol proposal. The IMS routes the SIP INVITE request to the Wirelessfuture PoC server (Participating) triggered on the PoC service indication and the PoC address.

2. **Inviting the team members.** The Wirelessfuture PoC server (Participating) identifies that the Prearranged PoC group 'Chatham Wanderers' is not hosted in this PoC server sends the request to the Wirelessfuture IMS. The Wirelessfuture IMS routes the request according to the routing principles to OtherDomain IMS. The SIP INVITE contains the Pre-arranged PoC group identity, PoC address of Paul's PoC client, PoC service indication, Wirelessfuture PoC server (Participating) selected media parameters, and Talk Burst Control Protocol proposal. The OtherDomain IMS routes the request to the OtherDomain PoC server (Controlling) based on the Pre-arranged PoC group identity.

3. **Team members alerted**. When the first ALERTING (SIP 180 Ringing) response is received the OtherDomain PoC server (Controlling function) sends the ringing provisional response towards Paul's PoC client.

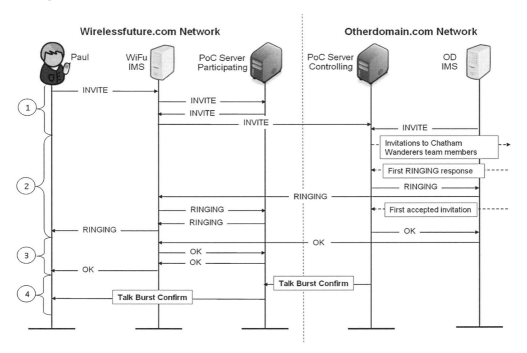

Figure 6.7 Signalling Flow for Pre-arranged Group Invite

4. **Pre-arranged session verified**. When the first PoC client accepts the Pre-arranged PoC session invitation, the OtherDomain PoC server sends an OK response to the Wirelessfuture PoC server (Participating) along the same signalling path. The SIP OK contains OtherDomian's PoC server (Controlling) selected media parameters and selected Talk Burst Control Protocol. The PoC server sends an OK response to the PoC client along the same signalling path.

5. **Media exchanged between team members**. OtherDomain PoC server (Controlling) sends the Talk Burst Granted response to Wirelessfuture PoC server (Participating). Wirelessfuture PoC server (Participating) transfers the Talk Burst Granted response to Paul's PoC client.

6.4.3 Rejoin ongoing Pre-arranged group session

In instances when John has not initiated the Chatham Wanderers session, he can drop out of the Pre-arranged group session and rejoin the still ongoing session. When John wants to rejoin a PoC session, his PoC client generates an initial SIP INVITE and sets the Request-URI to the PoC Session Identity and includes the Session Type URI-parameter indicating the PoC session type 'session=Pre-arranged'. The SIP INVITE will include a MIME SDP body. The signalling flow for rejoining is shown in Figure 6.8.

The basic procedures of rejoining a Pre-arranged session are:

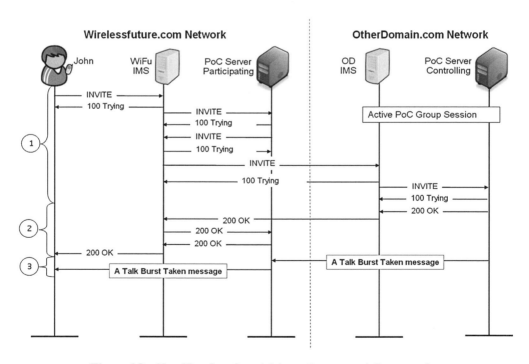

Figure 6.8   Signalling flow for rejoining a Pre-arranged Group session

1. **Rejoin the Pre-arranged session.** John wants to rejoin an ongoing Pre-arranged PoC group session and his PoC client sends SIP INVITE request to Wirelessfuture IMS. The SIP INVITE request includes the PoC Session Identity in the Request-URI. sip:PoC-SessionXYZ@poc.wirelessfuture.com; session=Pre-arranged. The SIP message contains:

```
INVITE sip:PoC-SessionXYZ@poc.wirlessfuture.com;
  session=Pre-arranged SIP/2.0
P-Preferred-Identity: 'John' <sip:John@poc.
  wirelessfuture.com>
Accept-Contact: *;+g.poc.talkburst; require;explicit
User-Agent:PoC-client/OMA1.0
Privacy: id
Contact: <sip: John@poc.wirelessfuture.com >; +g.poc.
  talkburst
Supported: timer
Session-Expires: 1800;refresher=uac
Allow:  INVITE,ACK,CANCEL,BYE,REFER,MESSAGE,
```

The Wirelessfuture IMS sends SIP 100 'Trying' to John's PoC client. Wirelessfuture IMS sends a SIP INVITE to the Wirelessfuture PoC server. The Wirelessfuture PoC server sends SIP 100 'Trying' to Wirelessfuture IMS. Using the PoC Session Identity the PoC server determines that the Pre-arranged PoC group session is owned by another PoC server and sends the SIP INVITE request to OtherDomain IMS:

```
INVITE sip: PoC-SessionXYZ@poc.otherdomain.com; session=
  Pre-arranged SIP/2.0
User-Agent: Server: PoC-serv/OMA1.0
Contact: <sip:PoC-SessionXYZ@poc.wirelessfuture.com>;
  +g.poc.talkburst
```

Based on the PoC Session Identity in the Request-URI, Wirelessfuture IMS sends the INVITE request to OtherDomain IMS.. OtherDomain IMS sends a SIP INVITE to Other-Domain PoC server. The OtherDomain PoC server responds with SIP 100 'Trying' to OtherDomain IMS

2. **Rejoin the Pre-arranged session is accepted.** The OtherDomain PoC server authorizes John's client to rejoin to the Pre-arranged PoC group session and sends SIP 200 'OK' response to OtherDomain IMS.

```
SIP/2.0 200 OK
P-Asserted-Identity:<sip:Chatham Wanderers@poc.
  otherdomain.com; session=Pre-arranged>
Contact: <sip:PoC-SessionXYZ@ poc.otherdomain.com;
  session=Pre-arranged>;
Supported: norefersub
```

The OtherDomain IMS sends SIP 200 'OK' to Wirelessfuture IMS:

```
SIP/2.0 200 OK
P-Asserted-Identity: <sip:Chatham Wanderers@poc.
  otherdomain.com;session=Pre-arranged>
Contact <sip: PoC-SessionXYZ@PoCServerX.networkX.
  net;session=Pre-arranged>;
```

Wirelessfuture IMS sends SIP 200 'OK' to Wirelessfuture PoC server and Wirelessfuture PoC server responds with SIP 200 OK to Wirelessfuture IMS. Wirelessfuture IMS sends SIP 200 'OK' to John's PoC client

3. **Media exchange between the Pre-arranged session participants.** OtherDomainPoC server sends a TBCP Talk Burst Taken message to Wirelessfuture PoC server, when another team member (PoC client) has permission to speak. Wirelessfuture.com PoC server sends the TBCP Talk Burst Taken message to the John's PoC client. (C). RTP media flows from the remote client to John's PoC client, through Otherdomain.com and Wirelessfuture.com PoC servers.

6.5 Chat Group Session

John is a part time taxi driver in the evening for a company called Fast Taxi. When a regular customer calls John, if he is not nearby John will use PoC to establish which taxi colleague is nearer to collect the regular customer. After a quick group discussion, John informs the regular customer that she will be collected in 2 minutes. The regular customer considers John to be her own taxi driver, since she is always collected with 2–5 minutes and is offered a 20% discount when using Fast Taxi. Group communication has benefited John, since he has more long distance customers and more scheduled drives. He also enjoys Chatting with the other taxi drivers when idling or waiting for customers. Most of the PoC traffic occurs around midnight. The Fast.Taxi Chat group is owned by the OtherDomain PoC server (Controlling function).

6.5.1 Join Chat Group Session

When John joins 'Fast.Taxi' Open Chat group, his PoC client generates an initial SIP INVITE and sets the Request-URI of the SIP INVITE request to the PoC Group Identity identifying the PoC group and the specify Session Type uri-parameter 'session=Chat'. The SIP INVITE includes a MIME SDP body that:

- sets the IP address and port number for the RTP session;
- includes the CODEC(s) and media parameters being offered by the PoC client for the PoC service;
- sets the IP address and port number to be used for RTCP at the PoC client;
- sets the Talk Burst parameter(s) and the port number(s) for the Talk Burst Control Protocol.

If John decides to be anonymous, the PoC client will insert value 'id' in the Privacy header.

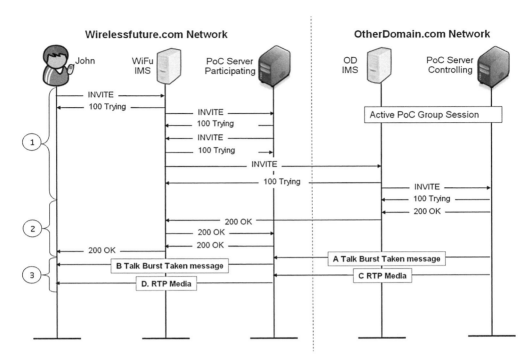

Figure 6.9 Signalling flow for joining a Chat session

As shown in Figure 6.8, the SIP INVITE is sent to the Wirelessfuture SIP/IP core (1) forwarded to the Wirelessfuture's PoC service (2) sent to OtherDomain.com SIP/IP Core (3) forwarded to OtherDomain.com PoC service. All other clients join the session (4). The high level signalling for joining the Chat group are shown in Figure 6.9.

The basic procedures of joining a Chat session are:

1. **PoC client attempts to join the Chat session.** John's PoC client sends a SIP INVITE request to Wirelessfuture IMS Core with Request-URI sip:Fast:taxi@poc.otherdomain. com; session=Chat. The SIP INVITE:

```
INVITE sip:Fast:taxi@poc.otherdomain.com; session=Chat
  SIP/2.0
P-Preferred-Identity: 'John' <sip:John@poc.
  wirelessfuture.com>
Accept-Contact: *;+g.poc.talkburst; require;explicit
User-Agent:PoC-client/OMA1.0
Privacy: id
Contact: <sip:John.poc.wirelessfuture.com>;+g.poc.
  talkburst
Supported: timer
Session-Expires: 1800;refresher=uac
Allow: INVITE,ACK,CANCEL,BYE,REFER,MESSAGE , SUBSCRIBE,
  NOTIFY,PUBLISH
```

Wirelessfuture IMS sends the SIP INVITE to Wirelessfuture PoC server. The Wirelessfuture PoC server responds with SIP 100 'Trying' to Wirelessfuture IMS. John's application to join the Chat group is routed to the correct network provider. Using the PoC Group Identity, Wirelessfuture PoC server determines that the Fast.Taxi Chat PoC group is owned by OtherDomains PoC server and sends the SIP INVITE request to OtherDomain IMS:

```
INVITE sip:Fast:taxi@poc.otherdomain.com; session=Chat
  SIP/2.0
User-Agent: Server: PoC-serv/OMA1.0
Contact:<sip:PoC-SessionXYZ@poc.wirlessfuture.com;
  +g.poc.talkburst
```

Based on the Chat PoC Group Identity in the Request-URI, Wirelessfuture sends the SIP INVITE request to the OtherDomain IMS. OtherDomain IMS sends SIP INVITE request to OtherDomain PoC server. OtherDomain PoC server responds with SIP 100 'Trying' to the OtherDomain IMS

2. **Application to join the Chat session is accepted.** OtherDomain's PoC server authorizes John to rejoin to the Chat PoC group session and sends the SIP 200 'OK' response to the Com IMS.

```
SIP/2.0 200 OK
P-Asserted-Identity: <sip: Fast.Taxi@poc.otherdomain.
  com;session=Chat>
Contact <sip:PoC-SessionXYZ@com.pcserver.otherdomain.
  com;session=Chat>;
Server: PoC-serv/OMA1.0
Supported: norefersub
```

Once John has joined the group, he'll be able to receive talk bursts from other Chat session participants. Each incoming talk burst is signalled with a 'TBCP Talk Burst Taken' message, followed by the RTP stream. Eventually, when John presses the PTT key, the Controlling PoC server authorizes John's talk burst with a TBCP Talk Burst Granted message, and John's voice will be distributed to all Chat session participants.

6.5.2 Group Participant Information

When John wants to receive information about other members participating in the Fast Taxi group Chat session, his PoC client subscribes to the Conference State Event Package by sending a SIP SUBSCRIBE request to obtain information of the status of a PoC session. His PoC client generates a SIP SUBSCRIBE request, uses a new SIP-dialog and sets the Request-URI of the SIP SUBSCRIBE request to PoC Session Identity or the PoC Group Identity.

The Controlling PoC function is the one managing subscriptions to the Conferencing Event package. When the subscription is successfully activated, it will start notifying John's PoC

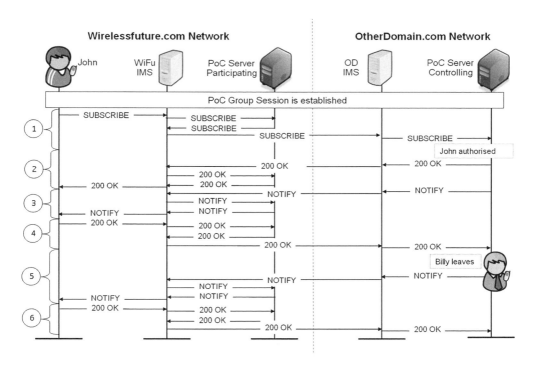

Figure 6.10   Signalling flow for group participation

client about status of PoC session participants during the duration of the session (e.g. active participants, new participants joining the session, participants leaving it). The high level signalling flow is shown in Figure 6.10.

The basic procedures of obtain group participation information are:

1. **Subscription for session participation information in an ongoing Chat session.** John's PoC client sends a SIP SUBSCRIBE request to Wirelessfuture's IMS. The SIP request includes the PoC Group Identity of the PoC group in the Request-URI and an Accept-Contact header with the PoC feature tag '+g.poc.talkburst'. The PoC client SIP message contains:

```
SUBSCRIBE sip:Fast.Taxi@poc.otherdomain.com SIP/2.0
P-Preferred-Identity: John' <sip:John@poc.wirelessfuture.
  com>
User-Agent:PoC-client/OMA1.0
Accept-Contact:*;+g.poc.talkburst; require;explicit
Contact:<sip:John@poc.wirelessfuture.com>
Event: conference
Subscription-State: active;expires=3600
```

The Wirelessfuture's IMS sends the SIP SUBSCRIBE to Wirelessfuture's PoC server. Wirelessfuture's PoC server performing the Participating PoC function does not recognize the PoC Group Identity as its own and sends the SIP SUBSCRIBE request to Wirelessfuture's IMS.

```
SUBSCRIBE sip:Fast.Taxi@poc.otherdomain.com SIP/2.0
P-Asserted-Identity: 'John' <sip:John@poc.wirelessfuture.
  com>
User-Agent: Server: PoC-serv/OMA1.0
Accept-Contact: *;+g.poc.talkburst; require;explicit
Contact: <sip:pocserver@ poc.wirlessfuture.com >
Event: conference
```

Wirelessfuture's IMS sends a SIP SUBSCRIBE request to OtherDomain's IMS based on the Request-URI.

OtherDomain's IMS sends the SIP SUBSCRIBE to OtherDomain's PoC server. Other-Domain's PoC server authorizes John as a group member to receive event information. The PoC group XML document associated to the Fast Taxi Chat group, stored in OtherDomain.com's PoC XDMS contains policy information that lets the PoC server decide whether to accept John's subscription. The authorization may be based on membership in the PoC Group, number of PoC users already subscribing to the event information or if the PoC user is a Participant in the ongoing PoC session. OtherDomain PoC Server sends SIP 200 'OK' to OtherDomain's IMS, which is forwarded to Wirelessfuture's IMS.

2. **Notification of participation is sent.** The PoC server performing the Controlling PoC function collects information about all Participants in the PoC session and sends a complete list of all Participants (referred to as a 'full' output in the conference state event package) in a SIP NOTIFY request to the OtherDomain IMS. The Request-URI sip:poc. server@poc.wirelessfuture.com. The SIP NOTIFY contains:

```
NOTIFY sip:John@poc.wirelessfuture.com SIP/2.0
Contact <sip:pocserver@poc.otherdomain.com >
Event: conference
Content-Type: application/conference-info+xml
Subscription-State: active;expires=3600
```

OtherDomain's IMS sends the SIP NOTIFY to Wirelessfuture's IMS. Upon receiving an incoming SIP NOTIFY request that is part of the same SIP dialog as the previously sent SIP SUBSCRIBE request John's PoC client displays the current state information of the PoC session or PoC group to the PoC user based on the information in the SIP

NOTIFY request body. The SIP NOTIFY contains an XML body such as the following one:

```
<?xml version='1.0' encoding='UTF-8'?>
<conference xmlns='urn:ietf:params:xml:ns:conference-info'
  xmlns:xsi='http://www.w3.org/2001/XMLSchemainstance'
  xsi:schemaLocation='urn:ietf:params:xml:ns:
conferenceinfo'>
  <conference-info entity='sip:Fast.Taxi@poc.otherdomain.com
    state='full' version='1'>
    <users>
      <user entity='sip:mary@otherdomain.com' state='full'>
        <display-text>Mary Quant</display-text>
        <endpoint entity='sip:mary@otherdomain.com'>
        <status>connected</status>
      </user>
      <user entity='sip:Billy.Hunt@otherdomain.com'
      state='full'>
        <display-text>Billy Hunt online</display-text>
        <endpoint entity='sip:Billy.Hunt@otherdomain.com'>
        <status>connected</status>
      </user>

      <user entity='sip:anonymous@otherdomain.com'
      state='full'>
        <display-text>Anonymous User</display-text>
        <endpoint entity='sip:anonymous@otherdomain.com'>
        <status>connected</status>
      </user>
    </user>
  </conference-info>
</conference>
```

3. **Participation information is accepted by the client.** SIP 200 OK responds are exchange between John's PoC client, the Wirelessfuture and OtherDomain IMSs and PoC servers.
4. **Participant leaves the session**. Billy leaves the Chat session. The OtherDomain PoC server sends a SIP NOTIFY to the OtherDomain IMS. This time only information about the Participant leaving the PoC session is sent (referred to by the conference state event package as a 'partial' output). The OtherDomain IMS sends a SIP NOTIFY to Wirelessfuture's IMS. This is eventually sent to the PoC client (John). The receipt of the SIP NOTIFY is acknowledged throughout the system.

   Hence, when a participant leaves the taxi session, John receives a SIP NOTIFY that contains an XML body such as:

```
<?xml version='1.0' encoding='UTF-8'?>
<conference xmlns='urn:ietf:params:xml:ns:
conference-info'
  xmlns:xsi='http://www.w3.org/2001/XMLSchemainstance'
  xsi:schemaLocation='urn:ietf:params:xml:ns:
conferenceinfo'>
  <conference-info entity='sip:Fast.Taxi@poc.otherdomain.
  com'
    state='partial' version='1'>
    <user entity='sip:Billy:Hunt@otherdomain.com'
    state='full'>
      <display-text>Billy Hunt here</display-text>
      <endpoint entity='sip: Billy.Hunt@otherdomain.com'>
      <status>disconnected</status>
    </user>
  </conference-info>
</conference>
```

6.6 Restricted Chat session example

Alice is a PoC dispatcher for VIP casino. She uses a desktop PC with a PoC client to coordinate the activities of teams who are responsible for the collection of important guests, welcoming guests to VIP rooms, and the general running of the casino. In her dispatcher role, Alice is a Distinguished Participant and can engage in 1-many-1 PoC group sessions with Ordinary Participants. She is also a PoC Group Administrator and has the authority to define, delete or modify group memberships from the Restricted Chat Group list.

To ensure that Alice has full control, the casino has defined the talk burst priority levels:

- **Preemptive priority**: Alice the dispatcher and Alf the casino owner have pre-emptive priority. A request to talk from the dispatcher (Alice) causes the current talk burst holder's permission to talk to be revoked immediately when it is received. When the talk burst is released or revoked, Alice the dispatcher who has requested to talk is granted the talk burst in preference to participants with high or normal priority.
- **High priority**: Managers and full time staff have high priority. When the talk burst is released or revoked, Managers and full time staff with high priority who have requested to talk are granted the talk burst in preference to participants with normal priority.
- **Normal priority**: Security staff has normal priority. When the talk burst is released or revoked, security staff with normal priority who have requested to talk are granted the talk burst only if there are no outstanding requests from Participants with higher priority
- **Listen only**. Cleaners and temporary staff have listen only priority. All requests to talk from Participants with listen-only priority are rejected.

All the PoC groups created by Alice are owned by the OtherDomain PoC server (Controlling function).

6.6.1 Group Advertisement

When Alice sends a Group Advertisement to the casino teams, the PoC client sends a SIP MESSAGE that includes an Accept-Contact header with the PoC feature tag '+g.poc.groupad' along with 'require' and 'explicit' parameters and PoC content in the form of MIME and.poc. group-advertisement–xml body. The basic flow is shown in Figure 6.11.

The basic procedures for Group Advertisement are:

1. **Sending the Group Advertisement to the recipients.** Alice's PoC client sends a SIP MESSAGE request to Wirelessfuture's IMS. The Request-URI includes the PoC group address. The Accept-Contact header includes the PoC feature tag '+g.poc.groupad'.

```
MESSAGE sip:VIPRoom2.Casino@poc.otherdomain.com SIP/2.0
P-Preferred-Identity: 'Alice' <sip:John@poc.
  wirelessfuture.com>
Accept-Contact:*;+g.poc.groupad; require;explicit
User-Agent:PoC-client/OMA1.0
Content-Type:application/vnd.poc.group-advertisement+xml
```

The xml body includes the ⟨Group⟩ element with the 'type' attribute set to 'dialed-in' because it is a Chat PoC Group. It also includes the ⟨uri⟩ element with the value set to the PoC Group Identity of the PoC group and sets the Request-URI according to the PoC

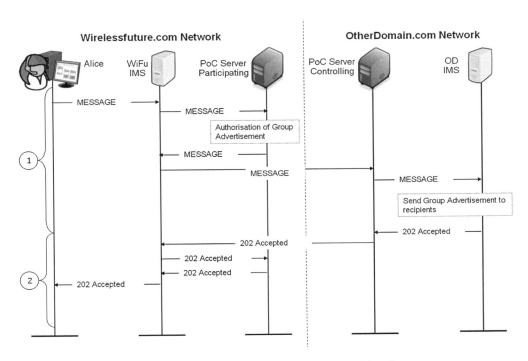

Figure 6.11 Signalling flow for sending a Group Advertisement

user's selection to a PoC address of a PoC user, or to a PoC Group Identity identifying a Chat PoC group or to an Exploder-URI identifying a SIP MESSAGE URI-list.

```
<?xml version='1.0' encoding='UTF-8'?>
<group-advertisement
  xmlns='urn:oma:params:xml:ns:poc:group-advertisement'
  xmlns:xsi='http://www.w3.org/2001/XMLSchema-instance'
  xsi:schemaLocation='urn:oma:params:xml:ns:poc:
  groupadvertisement'>
  <note>This is for those who work in VIP Room2</note>
  <group type='dialed-in'>
    <display-name>VIPRoom2.Casino</display-name>
    <uri>sip:VIPRoom2.Casino@poc.otherdomain.com </uri>
  </group>
</group-advertisement>
```

Wirelessfuture's IMS forwards the SIP MESSAGE request to the Wirelessfuture PoC server based on the PoC feature tag '–g.poc.groupad' in the Accept-Contact header. Wirelessfuture's PoC server authorizes Alice's PoC client to send a Group Advertisement and sends the SIP MESSAGE back to Wirelessfuture's IMS. Wirelessfuture's IMS sends SIP MESSAGE to OtherDomain.com's IMS using Request-URI sip:VIProom2.casino@ poc.otherdomain.com. The OtherDomain IMS sends the SIP MESSAGE to OtherDomain's IMS. OtherDomain's PoC server initiates the sending of the Group Advertisement to the recipients including Billy's PoC client.

2. **Group Advertisements are received by the recipients.** The PoC server sends SIP 202 'Accepted' to Wirelessfuture's IMS and is forwarded on to John's PoC client:

```
SIP/2.0 202 Accepted
P-Asserted-Identity: <sip:pocserver@poc.otherdomain.com>
Server: PoC-serv/OMA1.0 13.
```

6.7 Talk Burst Control Procedures without Queuing

In most group communication situations, floor control is adequately served without talk burst control queuing. Using the first come, first served model, the PoC session initiator is granted the first talk burst and there is no general requirement for controlled talk burst allocation because participants can access the floor as soon as it is vacant. This is particularly pertinent for the Ad-hoc PoC session, where group formation is limited to a short lifetime, such the duration of conference, and there is no need to set up talk burst priority levels.

It is expected that the majority of PoC servers and PoC clients will supports arbitration of Talk Burst Requests without queuing. The mechanism for Talk Burst Requests without queuing allows PoC server and PoC client to support:

- Talk Burst Request
- Talk Burst Granted
- Talk Burst Deny

- Talk Burst Release
- Talk Burst Idle
- Talk Burst Taken
- Talk Burst Revoke
- Talk Burst Connect, Talk Burst Disconnect and Talk Burst Acknowledgement messages.

6.7.1 Talk Burst Request Procedure at PoC Session Initialization

When John initiates the On Demand Ad-hoc PoC session with Mary, the permission to send a talk burst is granted to John's client. Figure 6.12 shows the Talk Burst Control flow for this scenario. The PoC session establishment request message from John's PoC client to the OtherDomain PoC server (Controlling) is called an implicit Talk Burst Request. When the Controlling PoC server has accepted the PoC session establishment, it acts as if it has received a Talk Burst Request. Note that initiating or joining a Chat PoC group session does not imply a Talk Burst Request.

The basic procedures for Talk Burst Request at session initialization are:

1. John has initiated a PoC session with OtherDomain.com's PoC server (Controlling). This creates an implicit Talk Burst Request and OtherDomain's Controlling PoC server sends Talk Burst Granted response message to John's PoC client. While this scenario shows the Talk Burst Confirm response message being sent after the PoC session is established with

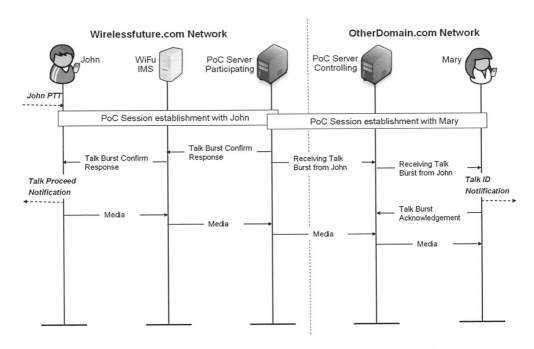

Figure 6.12 Talk Burst Request procedure at PoC session initialization

Mary, it can be transmitted also before or during the PoC session establishment with Mary depending on whether the PoC session indication is confirmed or unconfirmed.

2. At the same time, OtherDomain's PoC server (Controlling) sends a receiving Talk Burst Taken message to all other PoC clients in the PoC session (only Mary's PoC client is shown in the Figure). The received Talk Burst Taken message contains John's identity and the Mary's PoC client can display this identity for her.

3. In a case where acknowledgement is required, the receiving PoC client acknowledges the talk burst message. The Talk Burst Acknowledgement is received by the Wirelessfuture's PoC Server (participating) and is not sent to the OtherDomain's PoC Server (controlling).

4. When John receives the Talk Burst Granted response message, it provides a talk precede notification to John. John's PoC client then begins to send media to OtherDomain's PoC server (Controlling). OtherDomain's PoC server (Controlling) forwards this media to Mary and other PoC clients, if involved in the session.

6.7.2 Talk Burst Complete

Once John has introduced himself and topic of discussion, John releases the PTT Key and his PoC client sends a Talk Burst Complete message. Figure 6.13 shows the Talk Burst Control flow for this scenario.

The basic procedures for Talk Burst Complete are:

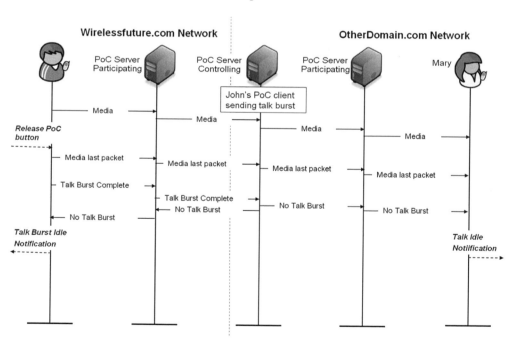

Figure 6.13   Talk Burst complete procedure

1. John has permission to send a talk burst. Media has been streaming from John's PoC client to OtherDomain's PoC server (Controlling) and this has been forwarding this media stream to the other PoC clients in the PoC session (only Mary is shown in the Figure).
2. John releases the PTT key and John's PoC client sends the last media packet to the Other-Domain PoC server (Controlling), which forwards it to Mary's PoC client.
3. John's PoC client then sends the Talk Burst Release message to the OtherDomain PoC server (Controlling).
4. After OtherDomain's PoC server (Controlling) has forwarded the last media packet, it then sends a Talk Burst Idle message to all Participants of the PoC session, including John's PoC client. Each of the PoC clients displays a Talk Burst Idle notification to its PoC user. Any Participant can now press the PTT key and request the right to speak.

6.7.3 Other Talk Burst Procedures

Talk Burst Control uses the ports (in the PoC client and PoC servers) negotiated at the SIP Session establishment. The PoC client and the PoC server support basic Talk Burst Control Protocol messages (Table 6.2).

6.8 Talk Burst Control Procedures with Queuing

In situations when group communication is coordinated there is a need to administrate priority levels to access the floor and Talk Burst Control queuing is required. This is relevant for enterprises in which managers and dispatchers have the right to barge onto the floor and start to give new task instructions. When the PoC clients and the PoC server (Controlling) in a PoC session support Talk Burst Control with queuing, the following procedures apply:

- Talk Burst Request with queued response
- Talk Burst Request cancellation
- Talk Burst complete with transfer to queued request
- Talk Burst stop with transfer to queued request
- Talk Burst Request with pre-emptive priority
- Talk Burst queue position request.

Talk Burst Control procedures with queuing places PoC clients in a queue according to a timestamp and priority. The PoC server (controlling) awards the PoC client at the head of the queue that has highest priority the permission to send the talk burst.

6.8.1 Talk Burst Request with pre-emptive priority

When Alice, the casino PoC dispatcher has an important announcement to make to the VIP room, she pushes the PTT key and is granted a talk burst, even though Billy is currently talking. In this example, Alice has pre-emptive priority to send the talk burst and Billy does not. Figure 6.14 shows the Talk Burst Control flow for this scenario.

Table 6.2   Talk Burst Messages

Messages	Explanation
TBCP Talk Burst Request	Used by the PoC client to request permission from the PoC server to send a talk burst
TBCP Talk Burst Granted	Used by the PoC server to notify the PoC client that it has been granted permission to send a talk burst. The Controlling PoC function includes information about the stop talking timer and can include the number of Participants in the PoC session at the time that this message is sent
TBCP Talk Burst Deny	Used by the PoC server to notify a PoC client that it has been denied permission to send a talk burst
TBCP Talk Burst Release	Used by the PoC client to notify the PoC server that it has completed sending the talk burst
TBCP Talk Burst Idle	Used by the PoC server to notify all PoC clients that no one has the permission to send a talk burst at the moment and that it may accept the TBCP Talk Burst Request message
TBCP Talk Burst Taken	Used by the PoC server to notify all PoC clients, except the PoC client that has been given permission to send a talk burst that another PoC client has been given permission to send a talk burst. NOTE 1: In the case of privacy the real identity of the PoC user, with the permission to send a talk burst, is replaced with an anonymous identity
TBCP Talk Burst Revoke	Used by the PoC server to revoke the media resource from a PoC client and can be used for preemption functionality, but is also be used by the system to prevent overly long use of the media resource
TBCP Talk Burst Acknowledgement	Used by the PoC client, when acknowledgement is required in the received TBCP message

When the PoC server and the PoC client support Pre-established sessions, the PoC client and the PoC server support the following additional TBCP Control message:

TBCP Disconnect	Used by the PoC server to close the PoC session using a Pre-established session while maintaining the Pre-established session
TBCP Connect	Used by the PoC server to notify all PoC clients using Pre-established session, that PoC session is connected

When a PoC server performing the Controlling PoC function and supporting queuing of TBCP Talk Burst Request messages, a PoC client supporting queuing of the TBCP Talk Burst Request message and a PoC server performing the Participating PoC function inserted in the media path supports the following additional Talk Burst Control Protocol messages:

TBCP Talk Burst Request Queue Status Request	Used by the PoC client to request the current queue position of a queued TBCP Talk Burst Request message
TBCP Talk Burst Request Queue Status Response	Used by the PoC server to notify the PoC client that the TBCP Talk Burst Request has been queued and is used to respond to a Talk Burst Queue Status Request message from the PoC client

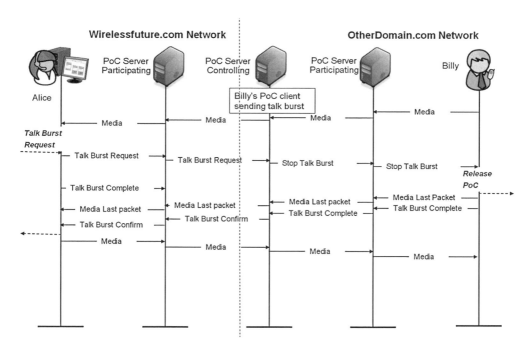

Figure 6.14   Procedure for Talk Burst Request with pre-emptive priority

The basic procedures for Talk Burst Request with pre-emptive priority are:

1. Alice the dispatcher is authorized to request permission to send a talk burst with pre-emptive priority does so, e.g. by pressing a PTT key, when another PoC user (Billy Hunt) who does not have pre-emptive priority has permission to send a talk burst. Alice sends a Talk Burst Request message to the OtherDomain PoC server (Controlling). The Talk Burst Request message identifies Alice as a PoC client requesting access with pre-emptive priority.

2. OtherDomain's PoC server (Controlling) determines that Billy Hunt does not have pre-emptive priority and that the pre-emption request from Alice is authorized and does not violate policies supported by OtherDomain's PoC server (Controlling), such as limits on the number of times or the amount of time that a PoC user is permitted to pre-empt other PoC users. OtherDomain's PoC server (Controlling) revokes permission to send a talk burst from Billy Hunt by sending a Stop Talk Burst message to Billy. The Stop Talk Burst message can indicate that Billy Hunt has been pre-empted by another PoC client. In this scenario, OtherDomain's PoC server (Controlling) grants Billy Hunt a grace period before revoking permission to send a talk burst.

3. Billy Hunt's PoC client sends a Talk Burst Permission Revoked notification to Billy. Billy indicates that he has finished speaking, e.g. by releasing the PTT key. Billy's PoC client sends the last media packet to OtherDomain's PoC server (Controlling) that is still forwarding the media to all other Participants of the PoC Session.

4. Billy's PoC client then sends the Talk Burst Complete message to the OtherDomain PoC server (Controlling).
5. After OtherDomain's PoC server (Controlling) has forwarded the last media packet from Billy's PoC client, it sends a Talk Burst Confirm response message to Alice's PoC client.
6. When Alice's PoC client receives the Talk Burst Confirm response message, it provides a Talk Precede notification to Alice the dispatcher. Alice's PoC client then begins to send media to OtherDomain's PoC server (Controlling). OtherDomain's PoC server (Controlling) forwards this media to the other PoC clients.
7. The first media packets forwarded to the other PoC clients are preceded by a Receiving Talk Burst message.

6.9 Summary and Conclusions

This chapter has given the reader a comprehensive overview of the most common signalling flows associated with group communication sessions. It has provided step-by-step message sequences for PoC Service Registration; PoC session set up – Ad-hoc group, Pre-Arranged and Chat; and for ancillary session actions such as adding a user and for inquiring who is else is listening in. Examples for talk burst procedures without and with queuing were also provided.

The chapter assumed that there was a PoC NNI Service Level Agreement between two operators. This assumption is critical for the successful commercial adoption of group communication service because it is unlikely that group members would belong to the same service provider. It is improbable that when strangers meet, like John meeting Mary at a remote conference, that they would both be registered to the same service provider. It is more realistic that a group of friends, such as Alf's team players would be made up of PoC users who have subscribed to two or more network operators.

Finally, the signalling examples are analogous to VoIP services that run on the Internet. This reminds us that multimedia group communication is more akin to computer networking than to Circuit Switched telephony.

6.10 References

[1] OMA Push-to-Talk over Cellular (PoCv1.0.1): 'PoC Architecture Document'; November 2006.
[2] OMA Push-to-Talk over Cellular (PoCv1.0.1): 'PoC Control Plane'; November 2006.
[3] OMA Push-to-Talk over Cellular (PoCv1.0.1): 'PoC User Plane'; November 2006.
[4] Rosenberg, J., Schulzrinne, H., Camarillo, G., Johnston, A., Peterson, J., Sparks, R., Handley, M. and E. Schooler: 'SIP: Session Initiation Protocol'. RFC 3261. IETF, June 2002.
[5] RFC 3267
[6] RFC 3558
[7] RFC 4867
[8] OMA Push-To-Talk over Cellular (PoCv1.0.1): 'PoC Requirements Document'; November 2006.
[9] RFC 4566: SDP
[10] RFC 3550: RTP
[11] OMA Push-To-Talk over Cellular (PoCv1.0.1): 'PoC XDM Specification'; November 2006.

7

Value Added PoC Services

7.1 Introduction

The evolution of multimedia group communications occurs in two directions. The first is to integrate the PoC and Presence solutions with existing Value Added service networks in order to drive up PoC adoption rates. Integrations of PoC with other IP technologies and messaging technologies are termed Value Added PoC services. The second direction is to follow OMA PoC 2, Presence 2 and XDM 2 specification and implement the new functionality. Version 2 of the PoC, Presence and XDM enablers are undergoing standardization with OMA, and with commercial products following in the coming years.

This chapter introduces the reader to Value Added PoC services opportunities. Section 7.2 infers that service providers can outsource their group communication administration to dedicated third parties. Section 7.3 gives examples on how PoC and instant audio functionality can be delivered by other Value Added Services (VAS) – SMS, MMS, email, IM and blogging. Section 7.4 describes the concept of PoC and infotainment channels. Section 7.5 combines Location Based Services with PoC and introduces the notion of condition based group communication.

7.2 Value Added PoC Service Roles

The service provider can increase the Return on Investment (ROI) of group communication by spreading the risk and engaging third parties to market group communication to professional PoC users or to provide infotainment content to basic PoC users.

7.2.1 Mobile Virtual Network Operator

A service provider can lease its group communication infrastructure to Mobile Virtual Network Operators (MVNO) and allow the MVNO to create small but lucrative business models. Group communication often requires personalized customer relationships and many operators do not have adequate sales resources to serve professional PoC users. Servicing small and intensive groups of professional PoC users is more suitable for a dedicated third party, such

Multimedia Group Communication. Andrew Rebeiro-Hargrave and David Viamonte Solé
© 2008 John Wiley & Sons, Ltd.

as an MVNO, who can combine multiple enablers – PoC, Presence and Location Based Services – and provide a single service to solve a variety of enterprise communication needs.

7.2.2 Value Added Service Provider

A service provider can add value to their PoC investment by allowing their PoC users to connect to a Value Added Service Provider (VASP) infotainment channels. An infotainment channel contains audio information on a wide range of consumable topics, from news, weather, and sports results, to tourist information. For PoC, a VASP is a content provider, who provides the different channels – news, sports, horoscope – and keeps the infotainment database up to date. The service provider allows the VASP to connect to their PoC service by using an NNI SIP interface and allocates a predefined data transmission capacity so as not to overload the radio network. The VASP can integrate their information channels with messaging applications (SMS and MMS) and with Location Based Services.

7.3 Integrating PoC Service with Existing Value Added Services

The main reason to integrate PoC and Presence with existing Value Added services, such as SMS, MMS, IM and email, is to reuse the immediacy and simplicity a PoC client offers to reach offline PoC users or end users who do not have a PoC subscription (Figure 7.1). In order to send a voice message to a non PoC user, all a PoC subscriber needs to do is select the contact and push the PTT key to record their voice and send the talk burst. A third party

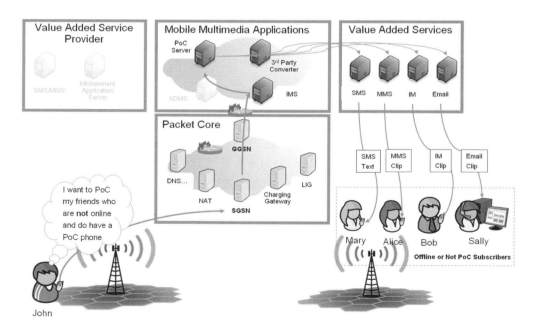

Figure 7.1   Extending PoC technology to reach non PoC users via existing Value Added Services

application gateway will discover the technology the target is using – MMS, IM, email or voice blog – and arrange the delivery. It is possible to send a PoC voice clip to a group of non PoC users by applying a user defined distribution list to direct the message to the targets.

7.3.1 PoC and SMS

The combination of PoC and a special SMS message to advertise a PoC group is useful for three conditions:

1. When the recipient is a PoC subscriber and is offline;
2. When the recipient is a PoC subscriber and is not an active user;
3. When the recipient is not a PoC subscriber and can be registered to the PoC service.

If a PoC Subscriber has not chosen to automatically register to the PoC service when they turn on their mobile phone, the receipt of an SMS Group Advertisement (via SMS) can stimulate the PoC subscriber to go online and activate the service. The same logic applies to subscribers who are not active users of the PoC service – the SMS based Group Advertisement influences a return to PoC behaviour.

The receipt of an SMS PoC Advertisement can also encourage a user to subscribe to the PoC and Presence service. If a group owner sends an SMS to a non-registered subscriber, the acceptance of the SMS Group Advert can trigger the device management service to send to the PoC service settings over the air to the recipient's mobile phone. Once the recipient accepts the PoC connection settings, then they become a PoC user.

7.3.1.1 PoC SMS Advertisement Example

This example indicates how SMS can be used to improve user experience and could be called a PoC advertisement. The idea is as follows: when a PoC enabled subscriber (Alice) tries to contact another mobile user (John) using PoC, if the called party is unavailable, the PoC server will send an advertisement SMS towards him. This way, the service provider promotes usage and increases the likelihood that the called party is available the next time that Alice tries to contact (PoC) him. In particular, the following use cases are facilitated by this mechanism:

- *Activation of the PoC client.* The service notifies John that Alice tried to contact him via PoC means and that he may switch on the PoC client to be reachable in the future. John may then activate the PoC client – in case it was off for some reason. This case applies to provisioned subscribers who *do* have a PoC enabled device, but are not using it at a given time.
- *Service provisioning and activation.* In this case, the service notifies John that Alice tried to PoC him. In particular, John may have a PoC capable client but he has never used the service, is not familiar with it and possibly the PoC settings are not activated. In this case, the SMS message may contain some explanation on how to request service provisioning, possibly including a reference to the fact that a well-known contact (Alice) has tried to

contact him using this new and attractive service. For Java and MIDP 2.0 compliant mobile phones, this mechanism could even provide a link to a downloadable Java PoC client, if available.

- *Service advertisement.* This is the simplest use case. The service simply notifies John that Alice tried to establish contact via PoC means. However, the recipient of the SMS message may or may not have a PoC enabled device. In a case where he does not have it, he is less likely to become reachable via PoC in the short term.

Observe that, in the first two cases, the likelihood that the called party is reachable at a later stage increases due to the *Service Advertisement* feature. This way, the user experience becomes progressively better, and service adoption grows in a so-called 'viral' way, as it is promoted from user to user.

This type of mechanism is particularly interesting from the point of view of operators. Compare the following two examples:

1. When Alice tries to PoC John, he will receive an SMS message containing the text: *'Hello John. Your friend Alice, +3583000000, invites you to join this new fancy service called Push-to-Talk. Send an SMS to 1414 and you will be able to enjoy it for free during a week.'*
2. A Customer Care agent calls John one evening and tells him: *'Dear Mr. John, I am calling in name of Operator XYZ to tell you about a fantastic service called Push-to-Talk . . .'*

It seems obvious that John will feel much more comfortable in the first case, in which the advertisement is triggered by a contact he probably trusts (Alice). He may even call her to comment about this new service (in which case Alice may even – and unnoticeably – become a product *prescriptor*).

Another drawback of customer care based advertising is the fact that enabling the PoC service for a single subscriber is useless: the operator needs to build a proposition so that other colleagues or members of John's family become subscribers of the PoC service at the same time. This requirement further complicates service provisioning and the design of the promotional campaigns. Observe that with the triggered *Service Advertisement* function, as soon as John receives Alice's advertisement and becomes a PoC subscriber, he will be able to start a PoC session with – at least – one PoC contact: Alice.

Some additional fine tuning and enhancement of this feature might include providing a polite recorded message to Alice (to ensure that user experience is not frustrating when the call could not be completed) or automatically notifying Alice when John becomes available for the PoC service in order to facilitate service usage. Figure 7.2 shows an example use case presenting how this feature could work.

In the picture above, when the PoC server detects that John is offline, it uses a dedicated interface with the SMS Center to send a notification to John (in this case John is a regular PoC subscriber, so the notification tells John that he may activate the PoC client). Some extra logic at the PoC server is required in addition to sending a '480 Temporarily Unavailable' SIP answer to Alice. An SMPP (SMS Messaging Peer-to-Peer Protocol) interface is used to connect the PoC server with the SMSC.

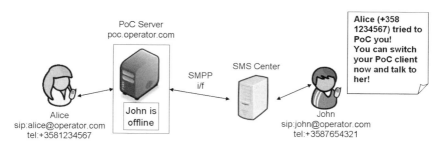

Figure 7.2 Offline PoC Advertisement scenario

7.3.2 Push-to-Voice Message

The combination of PoC and voice messaging extends the reach of PoC to non-PoC subscribers with the following benefits:

- PoC users can reach any mobile regardless of their messaging capabilities;
- PoC users can reach mobiles outside the service provider's domain;
- An application gateway stores and forwards the PoC talk bursts as audio message in a non real-time manner – the PoC talk burst is delivered when the mobile recipient is ready to accept it.

In general, there is still little penetration of PoC enabled terminals as many client vendors have not included PoC functionality in their handsets. However, many handsets are MMS capable and some can process audio messages. The integration of PoC and audio messaging circumvents this problem and enables PoC users to send their talk bursts as recorded voice notes to any contacts. The end user steps are quite simple:

1. John, a PoC subscriber, selects a contact from his phonebook;
2. John pushes PTT key to send a talk burst;
3. If that contact (Alice) does not have an active PoC client but does have MMS, an application gateway will convert that talk burst into a MMS message with a voice attachment and deliver it to the contact (Alice);
4. In response, the MMS user (Alice) can reply by recording a voice MMS and send it back to the PoC user.

To enable this service the operator will need an application gateway to automatically detect the terminal type and to deliver the voice message. The voice message will be delivered as:

- AMS (audio message) if the terminal is capable of receiving it;
- MMS, if the terminal is MMS enabled and not AMS enabled;
- SMS, with the note '*you have received Push-to-Talk message from Vladimir, call to listen to the message.*'

7.3.3 Push-to-IM

The combination of PoC and instant messaging extends the reach of PoC to non-PoC IM subscribers with the following benefits:

- PoC users can reach any online Internet instant messenger user;
- PoC users can reach any mobile regardless of capable of instant messaging;
- PoC users can reach mobiles outside the service provider's domain;
- An application gateway stores and forwards the PoC talk bursts as audio message in a non real-time manner – when the IM recipients ready to accept the recorded PoC talk burst.

The integration of PoC and IM has the added benefit of sending personalized audio messages to Internet users and therefore widening the reach of instant PoC communication. The end use steps are again quite simple:

1. John, a PoC subscriber, selects a contact (Mary) and sends a talk burst;
2. Mary has either set her preferences for communication to IM or is only available on IM;
3. The application gateway will transmit the talk burst as an audio message to the IM client, enabling Mary, the IM user, to hear the talk burst and reply to it via her IM client.

7.3.4 Push-to-Mail

The combination of PoC and email extends the reach of PoC to non-PoC subscribers with the following benefits:

- PoC users can reach any person with an email account, anywhere;
- PoC users can reach any mobile with push email capabilities;
- PoC users can reach mobiles with push email outside the service provider's domain;
- An application gateway stores and forwards the PoC talk burst as audio message in a non real-time manner – the email recipients can listen to the recorded PoC talk burst at any time.

The integration of PoC and email offers great potential to professional users, as well as for consumers, since there are times when users who are working in the field, such as at a construction site need to send a lengthy audio message back to their office for processing. A message, such as a subcontractor's verbal agreement to a deadline, can be recorded by the PoC application and delivered as an email to the sender's:

- Personal notes (deliver talk burst notes to user's personal inbox)
- Secretary (delivers notes to the secretary)
- Team (delivers talk burst to all the members of the team).

The end user steps are quite simple:

1. John, as a PoC user, selects individual or group recipients from his mobile PoC contact list;

2. John presses and holds the PTT key and speaks to record a message;
3. John presses the PTT key again to send the message;
4. Recipients, Alice and Mary receive the message in their email inbox, and can reply with a text message back to John's mobile phone.

The talk burst is recorded on the application gateway and sent to the appropriate email address or group of addresses. PoC users can have all of their sent messages stored also on the server, and can see when somebody has listened to a message. A third party may transcribe the talk burst into text. A more detailed example of Push-to-Mail is presented in the next section.

7.3.4.1 Push-to-Mail Example

This example shows how a corporate PoC application can include the Push-to-Mail solution. Consider the following scenario:

- A company has a number of distributed employees which use PoC regularly to communicate;
- Team managers are based in their offices. They want to stay in the loop of information whenever a relevant decision is made, but they do not like to be bothered by each and every talk burst exchanged between the members of their teams;
- In general, employees should not assume that their bosses are online and listening to the PoC phone. Sending a voice message via PoC is nice, but it is not enough. Having another reliable mechanism to deliver the same information would be an interesting added value to the basic PoC Service.

The above scenario is an ideal case for a Push-to-Mail solution. Let's imagine that the company deploys a Push-to-Mail server. Such an element could simply consist of a gateway that translates PoC calls into mail messages, which are sent to one or more mail accounts. From the PoC point of view, the Push-to-Mail server looks like any other PoC client (including its assigned PoC address).

The Push-to-Mail solution ensures that talk bursts are not only communicated via PoC means, but that they are also converted into mail messages which are safely delivered to one or more selected mail accounts. A mapping between one PoC address and one or more mail accounts is possible.

In the scenario described above, when field employees have relevant information they wish to communicate towards their team leader, they simply send a talk burst to a regular PoC address, which happens to be connected to the Push-to-Mail server. Observe that this configuration provides an additional degree of reliability to the PoC service: corporate users do not need to wait for the remote endpoint to become online, but can talk to the Push-to-Mail server and rely on the mail message to be read at a later stage.

Observe that the Push-to-Mail server will generally convert the voice talk burst into an audio file which is playable by a multimedia player application at the recipient's personal computer. Once the basic use case is understood, it is possible to enrich it with multiple additional extensions. To name a few we present the following examples:

- *Configuring Pre-arranged PoC groups that combine regular PoC clients with one or more Push-to-Mail addresses.* This configuration ensures both real-time (PoC) and reliable (mail) delivery. By combining both types of recipient in a single PoC group, instant and reliable delivery is achieved in a single transaction. Recipients get the talk burst via PoC, but in case they are offline or do not hear the talk burst properly, reliable delivery is ensured using the mail branch. In this case, for every human member of a Pre-arranged PoC group, two addresses should be included: the regular PoC address and the one connected to the Push-to-Mail service.
- *Combining Push-to-Mail with Presence.* By supporting Presence features, an additional degree of intelligence is achieved. A Push-to-Mail server may decide if it is necessary to forward a talk burst via SMTP or not, based on the Presence status of the original recipient: if the called party is offline, a mail message is generated, while this may not be the case if he or she is online.
- *Implementing a Mail-to-PoC mechanism.* By supporting text-to-speech features, a mail message could be converted into a talk burst and delivered to one or more mobile PoC users.

Figure 7.3 shows an example of how a Push-to-Mail solution could work.

In the above example, Charlie (sip:charlie@poc.operator.com) has information to share with the project manager (Mike). He sets-up a PoC session with the Project Management PoC group (sip:project.mgmt@poc.operator.com). This group comprises Mike (as a regular mobile PoC client) and a Push-to-Mail account (sip:mike.ptm@poc.operator.com). The talk burst sent by Charlie will be encapsulated in a multimedia file which is sent to a mail account using SMTP. A one-to-one mapping between sip:mike.ptm@poc.operator.com and mailto: mike@acme.com has been configured in the Push-to-Mail server.

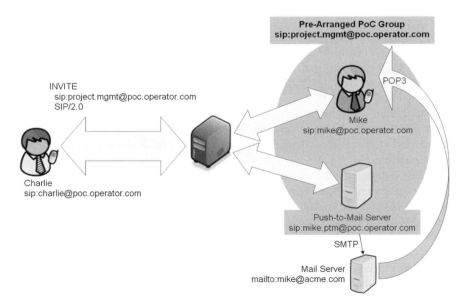

Figure 7.3 PoC & Push Email Solution

7.3.5 Push-to-Blog

The combination of PoC and posting voice blogs to the Internet users extends the reach of PoC to non-PoC subscribers with the following benefits:

- PoC users' voice blogs can reach online Internet users;
- PoC users' voice blogs can reach any mobile regardless with Internet capability;
- PoC users' voice blogs can reach mobiles outside the service providers domain;
- An application gateway stores and forwards the PoC talk bursts as audio message in a non real-time manner – when the voice blog listeners are ready to download recorded PoC talk burst.

The integration of PoC and voice blogging enhances the PoC service. The PoC users can share their recorded private thoughts or observations with selected individuals or make them available to the public. Voice blogs allow PoC users to interact with the larger Internet communities resulting in an increase of message flows and data usage for a service provider.

7.4 Push-to-Infotainment

An infotainment channel provides a fast effective method for PoC users to access information and entertainment over their mobile phones at the touch of a button. The always on nature of PoC over data networks provides to the PoC user with instant connection and rapid response to a variety of user queries. There are a wide variety of applications that can be delivered by using a third party PTT application platform.

The main benefits of infotainment channels are:

- PoC users do not need change any PoC client settings to access an infotainment channel – the only requirement for PoC users is the infortainment channel's address;
- PoC users can reach an infotainmnet channel at any time or place;
- PoC users do not feel inhibited about making an inquiry to a automated channel and therefore may use the service more regularly than calling a human;
- Service providers do not need change their current PoC infrastructure to be able to offer new information services.

Access to infotainment channels are the first step towards a Man-Machine PoC session. This type of Man-Machine PoC session relies on Voice Extensible Markup Language (VXML) to create audio dialogs between humans and automated systems. The audio dialog is based on key words. VXML is used to create voice Web portals, interactive voice response (IVR) and numerous other telephony applications, such as infotainment services.

Figure 7.4 shows an example infotainment setup with the service provider's Mobile Multimedia site being connected to a dedicated Value Added Service Provider. The service provider enables the PoC user, John, to access the infotainment channel at any place or time. The VASP hosts the Infotainment Application Server and provides up-to-date information to the information channels.

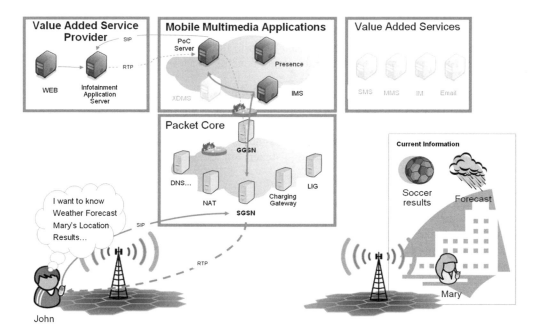

Figure 7.4   PoC Infotainment Application Platform Connection

7.4.1 Infotainment News Channels

An infotainment application platform can deliver news, quotes, weather or other web services data and forward these as a PoC voice feed to the end user. By selecting the topic of interest from the contact list, users can have the desired information spoken over their handset. Example applications that can be delivered based on such feeds or other data sources include:

- Push-to-Weather – Users have instant access to live weather information feeds for any location by simply speaking the name of the location. Additional or more detailed information can be obtained by making follow up requests;
- Push-to-Traffic – Users can obtain updated traffic information or road conditions based on highway number or location;
- Push-to-Headlines – Users can hear current news headlines or other information based on selected topics.

The end user steps to acquire a weather forecast are quite simple as shown in Figure 7.5.

7.4.1.1 PoC Infotainment Example

This example helps improve the ratio of successful sessions using infotainment use cases. In this case, as the called party is an operator-controlled PoC client, there is no issue in terms of unavailability of the end user or misconfiguration of the terminal. When the user accesses

Figure 7.5   Weather infotainment example

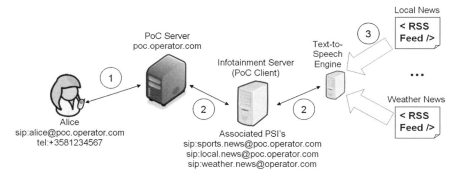

Figure 7.6   Infotainment set up

an infotainment service, he does not need to care about the remote endpoint's status and capabilities. For example, a news service may have a well-known Public Service Identity (PSI) such as: *sip:local.news@poc.operator.com*. In this case, local news is delivered to the customer, while sports news could be accessed at: *sip:sports.news@poc.operator.com*.

Figure 7.6 shows a potential implementation of such a service.

1. When a PoC subscriber presses the PTT key with the PoC address of the local news service (sip:local.news@poc.operator.com), the server redirects the SIP INVITE message towards the final PoC client, which happens to be the Infotainment Server.
2. The Infotainment Server processes the request and determines the suitable service. The PoC call is answered and connected to a Text-to-Speech (TTS) engine that delivers the news in form of packetized AMR packets over RTP, as any other regular PoC lient would do.

3. The Text-To-Speech engine is automatically fed with up to-date news that may come from an RSS service.

In addition to this described *pull* based service (i.e. the user requesting a news update), other variants may exist. For example, the user may install an alerting service that delivers up-to-date news every day at 8:30 AM. This way, a mobile subscriber may easily receive a news update when driving to their office, by enabling their mobile PoC phone in 'Automatic Answer' connected to a loudspeaker function (the user keeps their attention on the road and receives the news update without even having to touch the device).

7.5 Location Based Services with PoC and Presence

Value Added Service Providers or Mobile Virtual Network Operators can sell niche services that combine real time Location Based Services (LBS) with real-time mobile group communication to certain subscriber segments. These services would be complemented with Internet web sites and PC-based PoC clients. Different customer segments have different web client requirements:

- Professional PoC users – Corporate customers and government agencies with moving fleets require a secure Internet web client for professional tracking, fleet, workforce and asset management and PoC PC based client for group communication to send work task instructions;
- Basic PoC users who want to find friends, family members hobbyists and bloggers within a certain location require an easily accessible web site to subscribe to for privacy management, personal profile updates and to post pictures and messages, and a PC based PoC client for residential use;
- Businesses and tourist information boards interested in promoting Points of Interests (POI) within their municipality require a web client to update information on landmarks, shopping malls, restaurants, and to automatically send an associated infotainment channel address to PoC subscribers who are within a certain distance of the POI.

With access to LBS web clients and PoC clients, a PoC session can be automatically triggered based on different criteria.

The steps for LBS PoC service could be as follows:

1. Location Based Service Triggers based on:
   (a) Geofence (a predefined area)
   (b) Proximity
   (c) Matching Criteria
   (d) Schedule Alert.
2. Send an alert to the party that made the LBS request:
   (a) SMS text message giving location
   (b) MMS Map or Image giving location.
3. Start real-time PoC session with the located party or parties dependent on a predefined criteria:

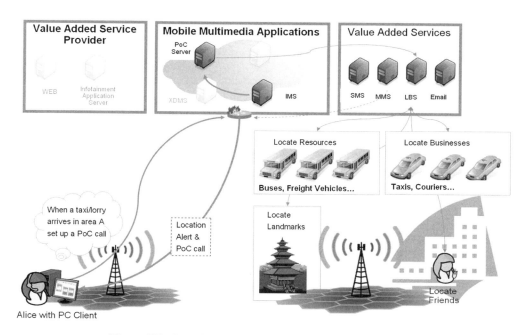

Figure 7.7 Location Based Criteria triggers with PoC session

(a) 1 to 1 session
(b) Ad-hoc group session
(c) Send an Group Advertisement to the open PoC Chat session
(d) Send an infotainment channel link.

A scheduled alert would find the location of participants and then invoke the PoC sessions.

Figure 7.7 shows how a work force coordinator could combine Location Based Services with PoC and Presence. Alice, the work force coordinator, knows which drivers are online but does not know their exact location. Instead of locating all vehicles, Alice has set a vehicle criterion: when a bus or taxi is near a given landmark, send a MMS map to the driver and set up a PoC session with the driver. If more than one driver arrives in the area, she can set up a group session. Table 7.1 links LBS triggers with PoC communication opportunities.

7.6 PoC PC Client Example

The idea of a PC based client is not to substitute the mobile device, but to let residential customers have a more powerful graphical environment that lets them easily configure their service, such as contact and group management, access list configurations and general application settings. The reason for such application is the fact that residential mobile users are generally non-technical, and they are not willing to use a limited device (in terms of screen size and lack of a QWERTY keyboard) such as a mobile phone, to perform complex management tasks like creating Pre-arranged groups.

Table 7.1 Location Based Services & PoC opportunities

Location Based Alert		Initiate Communication	Resultant Action
Geofence Alarms The LBS defines an area and request that when a certain terminal/ subscriber enters of leaves this area, an alert is triggered	*Send an SMS alert* if there is an intruder – an unregistered MSISDN – in the restricted area of the dock yard	PoC session with security guards	Investigate the marked location and make a report
	Send an SMS/MMS alert when my child leaves the school area	PoC session with my child	Collect the child from school
	Send an SMS/MMS alert if my car leaves this area	PoC session with an authority	Find out who has taken your car
Proximity Alert The LBS requests an alert to be raised when the distance between the locations of two terminals becomes smaller than a threshold	*Send an SMS/MMS alert* if any of my friends are near	PoC session with friends	Meet nearby friends for a coffee
	Send an SMS/MMS alert when I am near this landmark	Infotainment session about the landmark	Listen to history of the landmark
Matching alerts These are triggered when a terminal comes within a configured distance of a second terminal that matches a particular set of discrete criteria	*Match Me* with others who have registered to this game site	PoC session with those who match the criteria	Engage with game players
	Match Me with others bird watchers who have registered to this web site and are near me today	PoC session with those who match the criteria	Meet fellow hobbyists at this site to photograph the migrant birds
	Match Me with others who support my football team X and aged between 20–30 and are near me today	PoC session with those who match the criteria	Meet fellow supporters before the match
Scheduled alerts These are triggered periodically according to a configured time schedule	*Send an MMS Map* with location details of all resources (taxis, workers) in North part of the City at 14:00	Start a PoC session at 14:00	Allocate work tasks
	Send the location details of my friends in the city	Start PoC session at 19:00	Meet friends for evening out

On the other hand, some of these management tasks are required to facilitate service adoption (if a user is not able to create PoC groups, which are relevant for his communication needs, the service value he perceives is very poor). A way to combine a) easy mobile PoC usage, together with b) availability of management functions, is to let customers exploit a user-friendly PC based client that lets them easily configure service settings.

A PoC PC client may of course include not only group management features, but also the full set of PoC functionality, thus letting a subscriber decide which device he prefers to use in certain circumstances. The customer may use the PC client from home, while the mobile phone is the default device from any other location. In addition, service providers may decide to provide customers with a limited PoC client that lets them configure the service, but not initiate or receive calls (e.g. a PoC – XDM client that lets them configure Pre-arranged and Chat groups).

7.7 PoC for Vertical Segments

We have seen how location services are a key enabler for PoC to be successful in certain professional applications, such as transport or emergency services. If we go a step further we will see that the need to adapt and enrich PoC applies to other niche markets. In some segments, the ability of a mobile user to interact with a mobile device is very limited; in other cases, the handset is subject to extreme conditions and, in certain instances, the User Equipment itself must fulfil requirements such as being mounted in a vehicle such as a bus or a truck.

We will present below some simple use cases that show how PoC demands differ in certain specific corporate environments. Observe that as the market is further segmented, these needs become relevant only in very specific niche segments. In this case, the economies of scale of the more general PoC service cannot be applied, so the ability of PoC to become a truly competitive solution in these environments depends on its capability to really adapt to their needs, and provide real value to final corporate users.

7.7.1 Mary, The Truck Driver

Mary is a truck driver who uses PoC to communicate with her central office and to report incidents during her trips. She cannot afford losing her attention to the road for a single second, so all PoC operations should be available at a single click. When not driving, Mary uses the PoC device as her regular GSM phones to call her family.

Adapting PoC to the transport environment may require producing specific PoC devices, with at least one but possibly several dedicated buttons. Each button may represent a preconfigured operation, which lets Mary talk instantly to her central office without even having to look at the phone's screen. Being able to connect the phone to a sufficiently loud speaker is also a key feature.

7.7.2 Billy, The Bus Driver

In contrast to Mary, Billy is a bus driver. He drives the bus in a metropolitan area. The bus has built-in PoC capability, which lets Billy talk to the central office. Billy has his own GSM phone without any PoC capability.

In this case, dedicated PoC devices may even be built into a bus or a transport vehicle. Given its urban use, drivers do not need to have another device (apart from their personal phone, if they wish). The dedicated PoC client may offer several very basic operations by providing a very simple user interface based on large buttons and simple radio capabilities. In case the bus drivers are familiar with legacy trunking devices, the PoC user interface (e.g. buttons, microphone, speakers) mimics the legacy system, so that the technology upgrade towards PoC is almost transparent for end users.

7.7.3 Charlie, The Bricklayer

Charlie is a bricklayer working in the construction sector. He demands a PoC device that is robust and capable of working in difficult environmental conditions. It is key for him that he can use PoC without interfering with his regular job, so he should be able to talk and hear without even having to touch the mobile phone. Hence, a versatile and powerful headset system – possibly incorporating Voice Activity Detection (VAD) capabilities – is a capital requirement for Charlie.

7.8 Summary and Conclusions

The true value of the Push-to-Talk over Cellular (PoC) service is the simplicity of instantly reaching other online PoC users and sending them real-time audio messages. The added value PoC can add to other messaging services is to use the simple actions – PoC User selects a contact and PTTs – to send the audio as a voice recording to variety of terminating points: MMS, IM, email attachment or as an observation blog. Using the same actions, the PoC user can inquire and receive infotainment from information channels on a potentially large amount of sources.

Therefore, a PoC user can reuse the mobile phone as audio diary or personal note recorder and share his/her thoughts or directives with a multitude of end points. The PoC user can reuse the mobile phone as an information seeker – the latest weather forecast, sports results. Whilst weather and sports information is readily available on the Web it is not as an audio format and subscribers have to perform many clicks before getting the information.

The interactive nature of PoC and Presence can compliment other services, namely Location Based Services. Location Based Service gives up to date spatial data on the whereabouts of subscribers or resources, such as stolen vehicle, but does not enable communication to take action – who to report the stolen vehicle to? PoC compliments spatial awareness by bringing users together to solve such problem.

This chapter has focused on extending the audio capabilities of PoC service. The next chapter introduces PoC 2.0 new media types such as video and images. These media types can easily be added to Value Added Services and instead a PoC voice note being sent to the IM client or email group, a PoC video can be sent instead.

8

OMA PoC2 Group Communication Concepts

An interesting evolutionary route for multimedia group communication is to include peripheral machines within a group of mobile users. In this scenario, a group would consist of people who receive and share media streams from non human endpoints such as a surveillance camera and climatic monitoring equipment. The most fascinating element of the combination is the ratio between humans and machines – how many humans are required in the group? For example, the group can be highly personalized – *my camera, my mpeg player, my desktop and my mobile phone* – or it can be completely non-personalized – entirely made up of machines that work on a set of evaluation rules. OMA PoC2 does not focus on the issue of man and machines, but specifies the requirements for condition based sessions, condition reevaluation, PoC voting, and Man-Machine PoC sessions, which opens an avenue for automated, group decision making [1].

This chapter introduces the reader to selected concepts taken for OMA PoC2. OMA PoC2 is an evolution of OMA PoC1 and is fully backwards compatiable. Section 8.2 introduces new high level actors – lawful inception agencies, and non packetised switched P2T systems. Section 8.3 outlines possible use cases for condition triggered sessions and non real-time session control. Section 8.4 covers basic technical concepts related to different media types and simultaneous floor control.

8.1 Group Communication Roles

The group communication family extends with the introduction of the OMA PoC2. It is highly probable that, in the near future, service providers will have to copy in the lawful interception agency for warranted users. To extend the walkie-talkie reach in certain markets, it may be possible to convert OMA PoC media to a format understood by non packetised P2T networks.

8.1.1 Service Provider

The service provider or more commonly, the mobile network operator assumes the dominant role for multimedia group communication. The service provider owns or leases the infrastructure and has invested in:

Multimedia Group Communication. Andrew Rebeiro-Hargrave and David Viamonte Solé
© 2008 John Wiley & Sons, Ltd.

- IP/SIP Core and Application Servers to allow subscribers to form groups using clients and enable users to '*set up their own multimedia groups and exchange multimedia is real time*'. Unlike OMA PoC1, if a participant is absent from a group session, the participant will be able to download the exchanged media from a network repository.
- Radio environment that allows subscribers to quickly form mobile groups and '*exchange multimedia in groups in real time and be able to download multimedia after a session*', in which the participants are not physically visible to each other.

To ensure a successful group communication service, the service provider needs to ensure clear administration strategies:

- A marketing strategy that constantly informs their subscribers the general concepts of multimedia group communication, how to relate mobile distributing multimedia – PoC speech, audio, video, images, test to their lives in groups, and how to administrate the experience of sending and receiving multimedia;
- Attractive price strategy for group communication. Since group communication is predominately a sending/receiving service – *one person sends the other five participants' receive* – the aggregate price for receiving data should always be low;
- An excellent quality of service strategy – in which the call set up, is almost instant and the round trip time of real time voice and video from the speaker party to the group listeners & observers, regardless of the physical location, is the same as if the group was talking to each other in the same room.

8.1.2 Lawful Enforcement Agency

PoC2 defines a generic framework of PoC events which may be reported to a lawful enforcement agency [1, 2]. These elements include information such as participants of a session (including those who join it anonymously), information about sent and received talk bursts and timestamps of all reported events. Actual interception of the media traffic may require an additional interception rule at the bearer level (e.g. GPRS Legal Interception Gateway, LIG). That is: the PoC Lawful Interception function may determine the connectivity parameters of an RTP media stream (e.g. source and destination IP address and UDP ports) and mandate replication of that stream towards the law enforcement agency.

The main ingredients for lawful interception are:

- Lawful Interception authorization process;
- Identified PoC user whose PoC session have been lawfully authorized to be intercepted and delivered to a law enforcement agency;
- Law Enforcement Monitoring Facility in the service provider's network and a destination to send the results of the Lawful Interception.

8.1.3 P2T Network

This OMA POC2 feature enables a packet switched PoC service provider to interwork with private or public circuit switched or non SIP/IP packet switched networks that provide

Push-to-Talk services similar to PoC [2, 3]. The extension of packet switched PoC to circuit switched or remote Push-to-Talk (P2T) requires:

1. Address of external P2T network or remote network;
2. Address of P2T user and a logic that points the P2T address to an Interworking Function Agent that will convert signalling and media;
3. Interworking function to convert PoC SIP signalling, talk burst control & media burst control an to external P2T based session signalling talk burst control and media burst control;
4. A mechanism to charge for the PoC/P2T session.

Interworking with external P2T networks or non SIP/IP packet switched networks allows PoC users to set up sessions with CDMA PTT and propriety PTT systems. However, in practice the interworking will be complex to implement as the two systems have different call setup, round trip time values and – possibly – different types of user experiences.

8.1.4 PoC End Users

The end user's role is to subscribe to the service provider's multimedia group communication service. The subscriber uses client(s) to convey their group communication desires. For mobile group communication, the client is embedded on the mobile phone. For stationary group communication, client can be embedded on a PC or any other fixed device. OMA PoC2 introduces the notion of an autonomous PoC client being attached to a non-human endpoint such as a surveillance camera or climatic sensor. As an example, a PC client may invoke the PoC session and manage the autonomous transmission of video or machine data to an associated PoC group.

An important feature introduced by PoC2 is the capability to segment the customer base in different types of profiles. The service provider may offer a basic 'best-effort' PoC service to PoC basic users, while better quality is provided sequentially to PoC professional users. By assigning the right profile to each PoC session, a PoC server can properly determine how sessions should be prioritized under high load situations, or in which cases pre-emption can take place (e.g. a talk burst from a session having higher priority can interrupt a talk burst which was already being delivered).

By enabling different QoE profiles, the PoC service provider can design a service tailored to the needs of the end customer. These needs may range from basic best-effort communications among basic PoC users ('Basic' QoE profile) to pre-emptive priority ('Official Government Use' QoE profile) communications among officials managing emergencies, with 'Premium' and 'Professional' QoE profiles mapped in between.

New corporate, professional and residential features such as invocation of the PoC client from a web browser [4], moderator controlled sessions, upgrading to full duplex voice (circuit-switched or VoIP) or the ability for an authorized participant to drop participants from ongoing sessions will become possible with PoC2 as well.

8.1.4.1 Basic PoC Users

Basic PoC users are a service provider's subscribers who purchased the service to form group sessions for fun and entertainment, hobbies, college events, or for coordinating small businesses [5]. Basic PoC users form a PoC community. OMA PoC2 brings improved functionality to OMA PoC1.

- Invited Parties Identity Information – the PoC user is informed who has been invited to a group session;
- Manual Answer Override – the inviting PoC client can override the manual answer mode of the other PoC user(s) he/she is inviting to a PoC session;
- New media types – users can now share video or images while engaged into a PoC2 session;
- Converting PoC calls to full duplex calls – this allows the PoC session participant to request the other PoC session participants to set up another independent full duplex voice call (either circuit switched voice call or voice-over-IP call).

8.1.4.2 Professional PoC Users

Professional PoC users are enterprise subscribers who have purchased the service to coordinate their workforce, such as government departments, utilities and freight companies. These organizations may allocate PoC group administrator rights to define, delete or modify PoC group memberships. A professional PoC user can also be a Value Added Service Provider who specializes in providing group communication to niche industries [5].

Professional PoC users use role definitions to control group communication. A PoC dispatcher sends and receives media from fleet member(s). A PoC fleet member is similar to an Ordinary Participant and can only receive and send media to an allocated PoC dispatcher.

A PoC dispatcher has more functional rights to manipulate and control an ongoing PoC session that the Distinguished Participant outlined in PoC1. The PoC dispatcher can:

- Create groups of groups, thus enabling for easy combination of teams and Ad-hoc conversations, which should help organizations improve their productivity;
- Invoke PoC from a web browser and let users invoke the PoC client from a web page;
- Transfer PoC sessions thus maximizing flexibility and efficiency of the resources dedicated to the dispatch/coordination function:
  (a) Transfer PoC session from one set of workers to another;
  (b) Merge PoC sessions into a single PoC session;
  (c) Split PoC sessions to many PoC sessions;
- Moderate PoC sessions and act as a privileged user:
  (a) Expel participants from an ongoing PoC session;
  (b) Allocate talk priorities – pre-emptive priority to low priority;
  (c) Allocate Media Burst Control – to control a media according to predefined rules and procedures.

The following Figure (8.1) and Table 8.1 shows example roles involved in multimedia group communication. The service provider infrastructure simultaneously supports different types of PoC groups and related PoC sessions. The service provider is the root group administrator and allows the PoC community to form the different group types. The service provider comprises four services: SIP/IP Core (referred to as IMS), PoC server with a NW PoC Box,

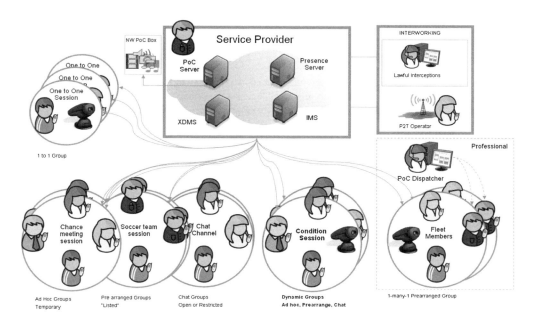

Figure 8.1   OMA PoC2 introduces new group functionality to the PoC community

Presence server and XDMS. The service provider's site multimedia operation is connected to a Lawful Interception Agency and to external P2T operator. There are five session types shown – 1 to 1; Ad-hoc session, Pre-arranged session, Chat session and Dynamic session. PoC users and non human endpoints, such as cameras are participants in a PoC session. For professional PoC, the participants may be subdivided into PoC dispatcher (session moderator) and fleet members.

8.2 Multimedia Group Communication Use Cases

It is the service provider's task to market multimedia group communication use cases to their subscribers. This task is even more strenuous than OMA PoC1, as there are not only many different types of groups (as shown in Figure 8.1): there are many different media types that can be transmitted to many different device types.

However, OMA PoC2 focuses on session dynamics and group communications, which can once again be simplified as a connection between a predefined set of users, in which:

- The set of users can be invited to a PoC session by another PoC user;
- The set of users can be invited to PoC session by a condition based automative PoC client;
- The set of users can join in to the session by their own free will – this is similar to joining a channel such as an Internet chat room;
- The set of users can join in to a prerecorded session by their own free will – this is similar to accessing an infotainment channel.

Table 8.1   Push-to-Talk over Cellular main communication roles (OMA PoC2)

Concept	OMA PoC Definition
External P2T Networks	Private or public circuit switched or packet switched network that provide Push-to-Talk services similar to PoC services
Remote PoC Network	Other PoC network or inter working function to external P2T network
Remote PoC Client	PoC client that resides on the User Equipment that supports the PoC service while accessing the PoC network via a potentially non IMS enabled SIP/IP based network
PoC Remote Access User	A user of the PoC service accessing the service potentially via a non IMS enabled SIP/IP based network, not necessarily using a PoC client – a PoC user, with a valid subscription, accessing PoC services via a PSTN terminal
P2T Address	A P2T address identifying a P2T user. The P2T address can be used by PoC users to communicate with P2T users. The P2T address used in a PoC network points to the PoC Interworking Agent of the P2T sser in the PoC Interworking service
P2T User	A P2T user is a user of the P2T service provided by an External P2T Network
Lawful Interception	The legal authorization process, and associated technical capabilities and activities of law enforcement agencies related to the timely interception of signalling and intent of wire, oral or electronic communications.
Law Enforcement Monitoring Facility	A law enforcement facility designated as the transmission destination for the results of a Lawful Interception
Identified PoC User	A PoC user whose PoC session has been lawfully authorized to be intercepted and delivered to a law enforcement agency
PoC Dispatcher	PoC dispatcher is a participant in a 1-many-1 PoC group session that sends media to all PoC fleet members and that receives media from any fleet member. The PoC dispatcher is an enhancement to the PoC Distinguished Participant
PoC Fleet Member	A Participant in a 1-many-1 PoC group session that is only able to send media to the PoC dispatcher and that likewise is only able to receive media from the PoC dispatcher

The use cases in the next sections concern Condition Based sessions, as this is an interesting novelty introduced by OMA PoC2.

8.2.1   Condition Based PoC Sessions

OMA PoC2 provides to the service provider an opportunity to create innovative and lucrative use cases for PoC users and at the same time extend the PoC community to contain a range of stationary PoC clients attached to non-human endpoints. The underlying denominator for such Man-Machine groups is Condition Based PoC sessions. In this concept, the invocation of Ad-hoc, Pre-arranged, or Chat sessions are based on an evaluation of a set of rules. Only those PoC participants who match the criteria are invited to the session or sent a link to join the session. There are many derivatives.

The criteria for triggering a Condition Based PoC session are quite extensive for example:

- An dynamic Ad-hoc session is triggered by Presence notification:
  ○ A particular user goes online;
  ○ A set of predefined users go online;
- A Pre-arranged PoC session is triggered by a sensor that responds to a stimulus and provides an electronic signal to a PoC client:
  ○ Changes in environment (heat, light or pressure) – relevant for factory groups;
  ○ State operations, a critical network device or agent surpasses a critical threshold (becomes faulty) – relevant for operations & maintenance groups;
  ○ Proximity signal – an intruder in the house or bird activates a streaming sensor;
- A dynamic Chat session is triggered by location alert when a certain number of PoC subscribers are near each other;
- A dynamic Ad-hoc session is triggered by a time element – at a certain time of the week.

Figure 8.2 shows an example Man-Machine PoC session with a surveillance camera connected to an autonomous PoC client. The surveillance camera is triggered when a bird trips the sensor and the autonomous PoC client sets up a Pre-arranged PoC bird watcher session. Two members of the group accept, whilst a third, John, is absent. John has set an agreement to store media for these sessions in the service provider's NW PoC Box. The PoC woodpecker video is routed to the watching Pre-arranged group PoC members and the NW PoC Box. John downloads the woodpecker video afterwards, the when he comes online.

8.2.2 Condition Re-Evaluation

The membership of the dynamic groups can be re-evaluated dependent on the level or seriousness of the condition.

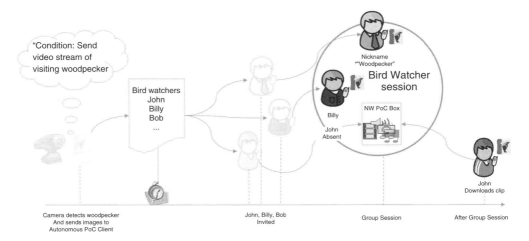

Figure 8.2   Dynamic Session – A non human entity (camera) invites bird watchers to a session

- When a particular PoC user goes, such as a Managing Director goes online, some low level employees are automatically removed from the PoC group discussion;
- If the sensor triggers a dangerous condition (exceeds the emergency threshold), then the group managers are invited to the session;
- If camera located in an off-limits area is triggered, then the dynamic Ad-hoc group members are re-evaluated according to an secret access list and non members of this list are removed;
- Re-evaluation of a condition can result in the creation of crisis sessions, where a condition is so urgent that certain PoC users are contacted – perhaps by an Interactive Voice Recorded (IVR) announcement system.

8.2.3 Crisis Handling

If the cellular and PoC service infrastructure are ready to support emergency services, the PoC Crisis Handling entity comes into play. This function – which can be seen as a privileged PoC client – is able to request sessions, which must be dealt with special treatment from the PoC service entity.

When a PoC Crisis Handling entity has been successfully authenticated it should be allowed to establish PoC crisis sessions with a higher degree of priority, reliability and reduced delay than regular sessions. Additionally, the privileged user calling from the Crisis Handling entity should be able to pre-empt other talk bursts and override the recipient's answer mode (i.e. use MAO if required).

Finally, the Crisis Handling entity may be able to record PoC crisis sessions, distribute discrete media (e.g. images) that can be relevant to manage the crisis situation, or implement the dynamic PoC group function, which can be used to set up a session to a selected group of emergency officers who are close to the crisis area.

All these features are shown in Figure 8.3.

A crisis handling scenario could look as follows:

1. A Crisis Event Reporter element (in this case an automatic alarm) may report a crisis at a given time. The alert reaches a Crisis Handling user (Rachel).
2. Rachel gets an immediate overview of the crisis event and the affected area.
3. Rachel sets up a session with a Pre-arranged PoC group, which involves three Crisis Handling Agents (Martha, Sam and Chris). Chris has Manual Answer mode configured, but the emergency session overrides this setting, so all agents' PoC clients automatically accept the incoming session, and the first talk burst spoken by Rachel is played in each loudspeaker.
4. Martha sets up another Crisis session, in this case using the Dynamic PoC group feature. She calls all regular PoC users located in the crisis area. Martha sends a pre-recorded message which warns all recipients that they should leave the affected area calmly, but immediately.
5. Martha uses PoC to share an image (received from the Automatic Crisis Reporter) with Chris, Sam and Martha. They use this information to better understand and decide how to tackle the issue.
6. Finally, Martha, Sam and Chris manage to sort the issue out. The fact that Mary, Brian and Charlie had left the area soon helped to minimize risks and ensure that the crisis agents could do their work in a more efficient way.

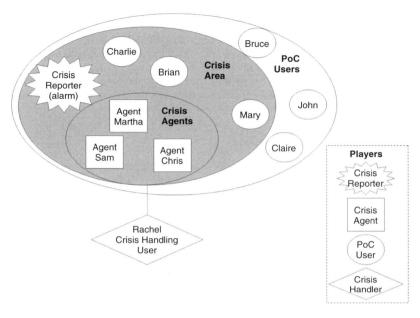

Figure 8.3   Crisis Handling Use Case

8.2.4 PoC Voting

Decision making is an important factor for group communication. The introduction of the PoC voting feature opens the way for some interesting scenarios, from simple agreements to remote democratic decisions.

There can be different methods:

- Open voting – All of the service provider's PoC community can vote on topical issues, such as opinion on a new fashion item;
- Closed voting – only group members can vote on an issue – such as a gaming group or a board group making a sensitive decision.

The results of the group poll can be:

- Disclosed – the result(s) are sent to all those who took part in the vote;
- Undisclosed – the result(s) are sent or kept by the originating client;
- Secret – the voters identity are not disclosed to the vote processing entity.

The results are sent to the participants:

- Real-time – the vote response to is forwarded to participants and can be updated;
- Accumulated – the aggregated vote response is forwarded to participants once only.

8.3 Multimedia Group Communication Implementation

The implementation of multimedia group communication – PoC, XDMS, and Presence with SIMPLE enablers – lies within the domain of computer networking and Internet Protocols (IP). OMA PoC2 increases the opportunity of connecting mobile PoC users with a range of stationary computerized devices. The expansion of RTP payload from talk bursts to rich multimedia payloads allows PoC users to hold conversations with other PoC users and simultaneously receive data from a high number of non-human endpoints.

8.3.1 The Concept of Media Burst

OMA PoC2 has defined media as forms of information exchanged between participants (Table 8.2). Media comes in different forms beyond half-duplex VoIP available in PoC 1. Media are referred to as Media Types. Media Types share a character of human perception and are distributed to group participants in real time (PoC speech) or in non real-time:

Table 8.2 Push-to-Talk over Cellular Group Types and Voting concepts (OMA PoC2)

Term	OMA Definition
Hierarchical PoC Group	A group within internal structure that is made up different levels.
Dynamic PoC Group	A Pre-arranged, Restricted Chat or Ad-hoc PoC group whose participants are restricted based on the evaluation of a set of rules.
Man-machine PoC Session	A PoC session between a PoC client with a human end user and another PoC client interacting with a non-human endpoint. Examples of non-human endpoint include: a software-controlled camera, a recorded announcement machine, or an application embedded in some other type of appliance.
Condition Re-evaluation	Repeated evaluation of conditions of PoC users against the rules that define a Dynamic PoC group. According to the evaluation results PoC users are invited to or removed from a PoC session involving the Dynamic PoC group.
Crisis Handling Request	A request where a PoC user needs the immediate attention of the invited PoC users.
Vote Processing Entity	Entity designated to process the voting result. This entity could either be the PoC server, the originating PoC client or a designated PoC client.
Vote Group Types	Open group vote: The voting is open to any PoC user (e.g. the unRestricted Chat PoC group). Closed group vote: The voting is restricted only to the PoC group members (e.g. the Restricted Chat PoC group).
Vote Result Types	Disclosed result vote: The voting result is sent to the PoC users who participated the voting. Undisclosed result vote: The voting result is kept/sent only by/to the designated PoC client (e.g. vote originating PoC client). Secret result vote: The voter's identity is not disclosed to the vote processing entity.
Vote Response Types	Real-time vote response: As and when vote response is received from a PoC client the response/accumulated response is forwarded to all the PoC clients. Accumulated vote response: The vote response from PoC clients is collected over a predefined time period. The voting result is computed and aggregate result is forwarded at the timeout. Any response received after the timeout is discarded.

- Audio (music)
- Video (with or without an audio component)
- Still image
- Text.

8.3.2 Multimedia and Time

The differentiation between real time and non real time media for PoC users is defined in terms of elements of time:

1. Continuous media with an inherent notion of time:
   - PoC speech
   - Audio (music)
   - Video.
2. Discrete media that does not contain an element of time:
   - Images
   - Text.

The introduction of time scales means that media does not have to been delivered in real time but can be stored and forwarded as a deferred messages similar to MMS or SMS messaging.

8.3.3 Deferred Messaging

Deferred Messaging is the OMA term for storing messages in a repository for later delivery. The ability to store multimedia for later replay has resulted in the creation of two new OMA PoC functional entities:

1. UE PoC Box is a function co-located with the PoC client in the User Equipment where PoC Session Data and PoC session control Data can be stored. The PoC Box can also replay recorded PoC Session Data to Participant(s) in a PoC session.
2. NW PoC Box is a functional entity in the PoC network where PoC Session Data and PoC Session Control Data can be stored. The NW PoC Box can also replay recorded PoC Session Data to Participant(s) in a PoC session.

8.3.4 PoC2 Signalling

OMA PoC2 signalling is performed by Session Initiation Protocol (SIP). The general flow of signalling for a PoC2 session is as follows:

1. A OMA PoC2 client (SIP User Agent) is embedded into the User Equipment.
2. A PoC Client registers to the service provider's PoC service via the SIP/IP Core (IMS) using SIP REGISTER.
3. The PoC client informs the PoC server about the client's multimedia capabilities and the PoC user's preferred settings to PoC service by sending SIP PUBLISH.

4. The PoC client invites other PoC users (identified by their corresponding PoC identities) to a PoC session or joins a PoC channel using a SIP INVITE
   a. SDP references the media content to be included in the PoC session
   b. SIP INVITE needs to account for offline PoC users who want to store the PoC session media in the NW Box.
5. PoC users leave the PoC session sending a SIP BYE.

8.3.4.1 OMA PoC2 Client Settings

Following sending a SIP REGISTration to the PoC service, the PoC client PUBLISHes the capability of the PoC client and the willingness of the PoC user to convey with related PoC client and PoC server functionalities. The PoC client settings include:

- Invited Parties Identity Information Mode
- Incoming Condition Based PoC Session Barring
- Incoming Media Barring
- Outgoing Condition Based PoC Session Barring
- Multiple Active Sessions.

The PoC client also sends the PoC user's own settings such showing an anonymous address to the other participants. Table 7.3 shows OMA PoC2 client settings.

Figure 8.4 shows a use case for multiple active sessions. An intruder has entered a secure area. An autonomous PoC client attached to the surveillance camera sets up a dynamic Ad-hoc session with a security team who are online. At the same time, John – the team leader – is alerted by the LBS system of the intrusion. John sets up an Ad-hoc session with the same online security team, therefore creating Multiple Active session.

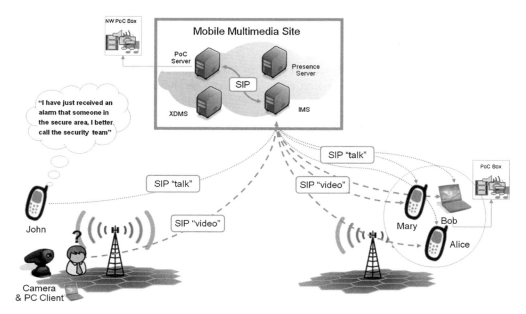

Figure 8.4 Multiple Active Sessions with a video media stream and audio media stream

8.3.4.2 OMA PoC2 New Media Types and CODECs Example

PoC2 supports new media types, apart from half-duplex narrowband audio. The following formats are supported:

- Still images (e.g. JPEG)
- Video (e.g. H.263 or MPEG4)
- Text messaging among groups of PoC users, based on the MSRP (Message Session Relay Protocol)
- File sharing in general (discrete media), provided that the required MIME types are supported by both clients.

Observe that the fact that PoC will support MSRP will represent an interesting add-on of messaging capabilities into the PoC service. Generalizing the PoC service offering we can see the PoC platform as a relay of generic traffic between two or more endpoints. From that point of view, one could think that this idea can grow up to cover a large number of use cases, more or less related to the initial PoC concept.

The addition of new media types has brought up new use cases for the support of simultaneous sessions. In particular, a PoC user may now support simultaneous sessions which may, in turn, involve one or more media streams (among which half duplex voice may only be one of the components). This situation leads to more complex simultaneous session support scenarios.

In OMA PoC1, the Participating PoC function is responsible for blocking an RTP media stream when it collides with another stream that is being delivered to the client device. OMA PoC2 optionally supports the so-called *Enhanced Simultaneous Sessions* (ESS) feature, which lets a PoC client receive all RTP streams, so that the PoC client is able to decide how to present them to the end user (e.g. discard a colliding stream, delay playback until the previous one is finished, discard selective media and always play audio streams).

Regardless of the ESS feature, PoC2 also has the capability – for UE not having SSS – to notify the user of a PoC Call Waiting, so that he or she can stop an ongoing session and switch towards the new incoming one, thus letting the subscriber be engaged into the most relevant PoC session at a given time.

In Figure 8.5 a comparison between SSS and ESS is shown. Let's assume that a client (supporting simultaneous sessions) is involved in three sessions at a time. There are two basic PoC sessions (only speech enabled) which will be governed by the SSS mechanism (the two sessions at the bottom of the diagram below). In this case, whenever there is a talk burst collision only the one from the primary session is delivered (so the 'crossed' talk burst is filtered out by the Participating PoC function). On the other hand, the device is involved in a third session (which comprises audio and video), and it has signalled the capability to support ESS for this one. In this case, the Participating PoC function delivers all RTP media streams (audio and video), and it is the device the one deciding how, when and if the received media is played towards the end user.

8.3.4.3 Enhanced invitation functions

OMA PoC2 supports new optional features that provide new functionality during the invitation stage. The use cases we comment briefly are the *Multimedia Invite* feature, the *Information About Invited Participants* feature and the *Alert for Unavailable PoC Users* function.

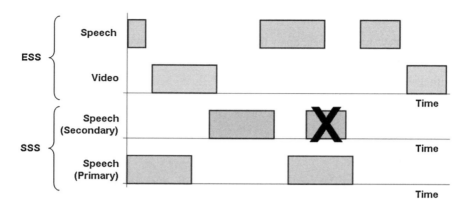

Figure 8.5 Support for Simultaneous Sessions and Enhanced Simultaneous Sessions

The Multimedia Invite feature lets an inviting client incorporate multimedia content to a SIP INVITE message, when initiating a PoC session. This feature can be used, for example, to deliver a small picture together with the signalling message, so that the callee can receive upfront information about the new session, or about its originator. This feature can also be used into Group Advertisement messages or IPA requests. The following message presents a potential use of the Multimedia Invite feature [6]:

```
INVITE sip:PoCConferenceFactoryURI.networkA.net SIP/2.0
Via: SIP/2.0/UDP pc33.atlanta.com;branch=z9hG4bK776asdhds
Max-Forwards: 70
From: 'PoC User A' <sip:PoC-UserA@networkA.net>
To: Bob <sip:bob@biloxi.com>
Call-ID: a84b4c76e66710@pc33.atlanta.com
CSeq: 314159 INVITE
Content-Type: application/sdp
Content-Length: (...)
User-Agent: PoC-client/OMA2.0 EStreet-Asbury1972/v0.99
P-Preferred-Identity: 'PoC User A' <sip:PoC-UserA@networkA.
  net>
Accept-Contact: *;+g.poc.talkburst; require;explicit
Alert-Info: <http://www.example.com/ringtone/cool-ringtone.3gp>
Contact: <sip: PoC-ClientA.networkA.net>;+g.poc.talkburst
Supported: timer
Session-Expires: 1800;refresher=uac
Allow: INVITE,ACK,CANCEL,BYE,REFER,
Content-Type: multipart/mixed
Require: recipient-list-invite
(...)
```

Table 8.3 Push-to-Talk Over Cellular Service Settings (OMA PoC2)

New Client PoC Settings	OMA Explanation
Outgoing Condition Based PoC Session Barring	A PoC service setting for the PoC client that conveys the PoC user's desire or the PoC service to block a particular outgoing PoC session request or PoC session join request based on a condition defined for outgoing PoC sessions
Multiple Active Session	An enhancement of Simultaneous PoC session and allows a PoC client to have several PoC sessions active at the same time without filtering of media
Invited Parties Identity Information Mode	A PoC service setting for the PoC server that conveys that the PoC client is able and PoC user is willing to receive invited parties identity information
Incoming Condition Based PoC Session Barring	A PoC service setting for the PoC client that conveys the PoC user's desire for the PoC service to block a particular incoming PoC session request based on conditions defined for Incoming PoC sessions.
Incoming Media Barring	A PoC service setting for the PoC client that conveys the PoC user's desire for the PoC service to block a particular incoming media type
New PoC Client Capabilities	
Receive Media Processing capability	The capability of the PoC client to handle media received from the PoC server
Transmit Media Buffering	A PoC client mode of operation where the PoC client buffers media in a buffer in the PoC client prior to the PoC server instructing the PoC client to transmit the media
Media Time Compression	A PoC client operation on media data to be transmitted, which compresses the media in time such that the compressed media data will be played out in a shorter time duration than the original uncompressed media data
New PoC Client Functionality	
UE PoC Box	A function co-located with the PoC client in the User Equipment where PoC Session Data and PoC Session Control Data can be stored. The PoC Box can also replay recorded PoC Session Data to Participant(s) in a PoC session

In the above case the inviting user suggests a cool ring tone to be played by the callee's handset, when receiving the multimedia INVITE message. The ring tone can be downloaded from the URL provided in the Alert-Info header. Observe presence of the 'User-Agent' header with the 'OMA2.0' string in it.

The 'Invited Parties Identity information' extension lets an OMA PoC2 user who is being invited to participate in a session, automatically receive information of the invited participants, even before he accepts to join the session. This way, privacy and security are enhanced, since the user will most likely know who he is going to talk to, if he accepts the invitation. This feature is actually enforced by PoC2, so that servers supporting it are recommended to use it. Although it does not completely remove the risk of users talking to unknown participants (since users are still able to join a session with an anonymous id), it enhances the degree of control that invited participants will have, thus improving service credibility and reducing its intrusiveness.

Two other interesting features introduced by OMA PoC2 are, respectively [1, 6]:

(a) The capability of ringing different PoC devices, which may be registered into the service from the same PoC identity (e.g. a user supporting PoC access from her mobile phone and from a desktop PC application). Each invited PoC device may use its own set of PoC settings and capabilities.

(b) The ability to notify an unregistered, unavailable or unprovisioned subscriber that another user tried to contact her via PoC means, so that she may actually activate the PoC client in her device and initiate a session against the original caller. This feature promotes service usage in a viral way. It can be used to warn users that long-lasting barring settings are preventing them from being contacted by other PoC users.

8.3.5 PoC Multimedia

PoC multimedia will be transmitted in RTP packets. In PoC2, RTCP carries Media Burst Control Protocol (MBCP) messages that are used to arbitrate requests from PoC clients for the right to send media (as opposed to TBCP used for audio only streams). With the introduction of multimedia, OMA PoC2 clients will need to [1]:

• Receive multimedia from the PoC server;
• Buffer multimedia as it waits for the PoC server to instruct the client to transmit the media;
• Compress media so that the data is sent efficiently over the wireless interface.

The UE PoC Box may also support storage of multimedia traffic when the user is offline.

The flow of PoC multimedia from a PoC client is known as a media burst.

The general flow for exchanging media bursts between participants is as follows:

1. SIP establishment phase:
   (a) SIP INVITE with SDP with referenced media parameters;
2. Inviting PoC user is granted first media burst:
   (a) Inviting PoC user receives MBCP Granted;
   (b) Invited PoC users receive MBCP Taken;
3. Inviting PoC user transmits a media burst:
   (a) PoC multimedia sent in RTP payload;
4. Invited PoC users receive the media burst;
5. Media Floor Control Entity ensures that only one PoC participant can access the media resource at the same time;
6. After a period of inactivity the RTP session is closed.

The RTP session dynamics are similar to OMA PoC1 talk burst flow. However, in OMA PoC2, the Media Floor Control Entity can handle one or more media streams, according to the session establishment negotiation. This means that there can be more than one floor control sessions for the participants to simultaneously engage in – *where the participants all watch the same video stream from a camera (one floor control) and the participants can comment on top of the video stream (a second floor control)* [2].

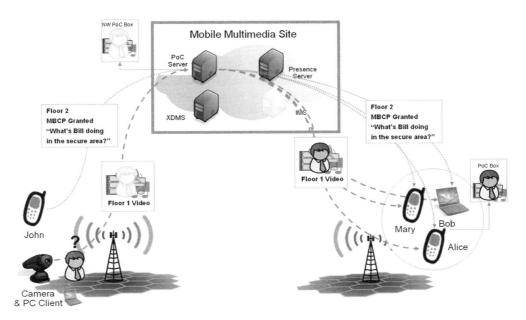

Figure 8.6 Two Media Streams Mediated by A Media-Floor Control Entity

Figure 8.6 shows a use case for Media-Floor Control Entity. Video of the intruder is streamed to the security team and is saved both in dedicated NW PoC Box and on Bob's local PoC Client Box (floor 1). At the same time, John is commenting on top of the video and instructs the security guards to take action (floor 2).

8.4   Summary and Conclusions

OMA PoC2 (Table 8.4) is undergoing specification and much will change in the next few years before there are commercial products. However, it is useful to ascertain the general direction PoC will go in the foreseeable future.

OMA PoC2 adds new actors to the high level concept: lawful interception agencies and non packet switched P2T networks. There is a greater focus on the professional PoC user than the basic PoC user. The professional PoC dispatcher is granted more opportunities to set up and modify PoC sessions.

Execution of PoC session takes a new direction – sessions can be set up 'on behalf' of the PoC groups by automated conditions and session may contain non human endpoints. Automating PoC sessions and holding conversations with non human endpoints may read as a little far fetched, however, as infotainment channels have showed, text to speech techniques and VXML have improved dramatically in the past few years.

The main difference between OMA PoC1 and OMA PoC2 is the introduction of new media – video, images and text to the group members. Media streams are sent in real time but can be received in real time or downloaded in non real time by the group members. The introduction of simultaneous media streams – speech and video – calls for a multiple floor control mechanism, so that users may be able to comment over live video footage.

Table 8.4   Push-to-Talk over Cellular User Plane Concepts (OMA PoC2)

Concept	OMA PoC Definition
Conversation	A conversation is a series of media bursts within a PoC session in which the inter-arrival spacing of the media bursts is less than a defined time interval; typically, the media bursts are associated with a logical exchange between two or more PoC users
Continuous Media	Media with an inherent notion of time (e.g. PoC speech, audio and video)
Discrete Media	Media that itself does not contain an element of time (e.g. images, text)
Audio	General communication of sound with the exception of PoC speech
Video	Communication of live-streamed pictures without any audio component
Media Burst	Flow of media from a PoC client that has the permission to send media to receiving PoC client(s)
Media Burst Control Protocol	A protocol for performing Media Burst Control
Media Burst Control Schemes	Way of using Media Burst Control and/or Talk Burst Control according to predefined rules and procedures
Media Burst Request Permission Level	A level of permission, which can be used to limit the PoC clients to request media burst
Media-Floor Control Entity	A Media Control resource shared by participants in a PoC session. The Media-Floor Control Entity is controlled by a state machine to ensure that only one Participant can access the media resource at the same time. One Media-Floor Control Entity can handle one or more media streams according to negotiation
Media Stream	A media stream is an instance of the transmission of a media type which is used as the basic unit to distinguish each media flow. Multiple media streams can be combined to transmit multimedia.

8.5 References

[1] OMA: 'Push-to-Talk over Cellular 2 – PoC Requirements', Draft Version (Work In Progress), August 2007.[2] Open Mobile Alliance: 'Push-to-Talk over Cellular 2 – PoC Architecture', Draft Version (Work In Progress), August 2007.

[3] OMA: 'Push-to-Talk over Cellular 2 – PoC Interworking Service', Draft Version (Work In Progress), August 2007.

[4] OMA: 'Push-to-Talk over Cellular 2 – PoC Invocation Descriptor', Draft Version (Work In Progress), August 2007.

[5] OMA: 'Push-to-Talk over Cellular 2 – PoC System Description', Draft Version (Work In Progress), August 2007.

[6] OMA: 'Push-to-Talk over Cellular 2 – PoC Control Plane', Draft Version (Work In Progress), August 2007.

9

Multimedia Group Communication Evolution: PoC2, XDM2, Presence 2 and Simple IM

9.1 Introduction

In this chapter we will focus on and describe the expected evolution of the next generation of SIP/SIMPLE-based OMA enablers, namely: OMA Presence v2, OMA XDMv2 and OMA SIMPLE Messaging v1. Version 1 of SIMPLE Messaging standardization timeframe is quite aligned with version 2 of PoC, Presence and XDM.

Additionally, after the most important use cases and features of PoC2 have been presented in chapter 8, we will devote a section in this chapter to describe PoC2 evolution from an architectural perspective.

The architectural work described in this chapter is based on the status of standardization activities at OMA at the time of writing (September 2007). Such work is at a reasonably advanced stage, with XDMv2 and SIMPLE IMv1 already having reached the *Candidate Enabler* status, and PoCv2 and Presence v2 expected to follow shortly[1].

It is worth observing that OMA initiated work around the topics described in this chapter between late 2004 and early 2005. Hence, a significant amount of work has been completed already, including requirements, architecture and technical specifications in all areas. We therefore expect the next sections to be quite close to reality, rather than simple speculation about potential solutions. In any case, significant evolution until all these enablers get eventually approved will take place. Hence, the reader should expect some misalignments between the contents of this chapter and the way the finally approved documents will look like. We believe, however, that general trends and principles of current specification work are captured here, and will certainly be of use to the advanced reader that wishes to get into deeper details

---

[1] However, the Candidate Enabler specification may suffer relevant changes until it reaches the *Approved Enabler* status. Interoperability work to be performed in upcoming OMA testfests is expected to uncover potential incoherencies or errors in the specification.

Multimedia Group Communication. Andrew Rebeiro-Hargrave and David Viamonte Solé
© 2008 John Wiley & Sons, Ltd.

about the evolution of future SIP services defined by OMA, and their impact on the end user's everyday activities in years to come.

9.2 Architectural Elements of OMA PoC2

9.2.1 Introduction

Since OMA PoC2 features and use cases have been described in the previous chapter, in this section we will focus on describing the evolution of OMA PoC2 from an architectural point of view.

It is worth noting that work in PoC2 has progressed substantially since the conception of the initial requirements document in early 2005, until the more mature – yet still not approved– technical specifications which are available at the time of writing (September 2007) [1]. In particular, during standardization work it was decided to split PoC2 standardization into two releases, namely: PoCv2.0 and PoCv2.1, with initial focus on the definition of PoCv2.0. In this section we will not cover in detail such split of functionalities, but present a high level overview of the overall version 2 Release of the PoC enabler.

Figure 9.1 shows a high level overview of the expected architectural evolution of the current enabler towards OMA PoC v2.0 and, eventually, towards PoC v2.1. New interfaces not available in PoC1 are marked in bold.

OMA PoC2 architecture is backward compatible with OMA PoC1. In particular, it shall be possible that a PoC1 device interoperates seamlessly with PoC2 infrastructure and vice-versa.

However, one of the changes that the reader may have directly observed in Figure 9.1 is the absence of the PoC XDMS. Effectively, a decision was made at OMA to concentrate all XDMS functionality into the Shared XDMS entity defined in XDMSv2. The reason for this architectural change is the fact that using correct XCAP tree management and correct XML namespace declarations, it is possible to host all XML documents into a single XDMS logical

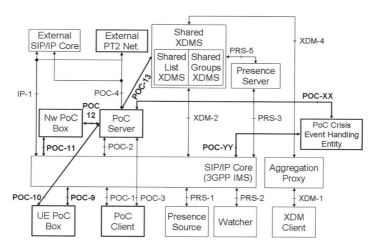

Figure 9.1 OMA PoC2 general architecture

entity. An XCAP compliant XDMS should be able to support a number of different AUIDs (including management of service–specific XML documents) with little effort. An enabler specific XDMS would only be defined in case a strong architectural need for that is identified. In the particular case of PoC2 such need was not detected. Regardless, XDMv2 should provide to an XDMv1 compliant client the same functionality that the PoC XDMS provides (hence, the Aggregation Proxy should hide to the client the fact that PoC specific documents are stored in fact in the Shared XDMS).

In any case, observe that the fact that a single *logical* Shared XDMS is defined in XMDv2 does not preclude that several *physical* XDMSs are used to ensure reliability and load balancing.

Apart from the multitude of functionalities introduced by PoC2 (as described in chapter 8), an important new element in PoC2 is the definition of a new *Media Burst Control Protocol* (MBCP). The idea is that the mechanism to arbitrate the right to send media has to cover not only voice talkbursts, but also new media introduced in PoC2 (e.g. text, images). Hence, a more generic protocol is defined in PoC2 [3], as a superset of the TBCP defined in PoC1. Furthermore, in future versions of the enabler we expect that OMA floor control protocol may adopt an even more generic mechanism, such as IETF-defined *Binary Floor Control Protocol* (BFCP) [4].

Observe that OMA PoC2 keeps the interfaces already defined in version 1 of the enabler. In addition, four new network elements have been depicted: the *UE-based PoC Box*, the *Network-based PoC Box*, the *External Push-to-Talk (P2T) network* and a *PoC Crisis Event Handling entity*. Observe the additional split of the Shared XDMS into the Shared List XDMS and the Shared Groups XDMS sub-modules: these new modules are described in further detail in section 9.3, while the PoC specific elements and interfaces are briefly presented below.

9.2.2 OMA PoC Box

The *PoC Box* provides session recording functions, in a similar fashion as the voice mailbox already available for voice services [1, 2]:

- The Network-based PoC Box can handle incoming sessions for a user who is temporarily out of coverage or unavailable.
- The UE PoC Box acts as a PoC mailbox for incoming talkbursts directly at the end user device. This way, if the PoC user does not wish to be bothered by incoming talkbursts (e.g. while in a meeting), these will be stored in her device, and she will be able to inmediately retrieve them at her convenience afterwards.

In the Network-based PoC Box case, the user must be able to retrieve stored sessions and talk bursts by either setting up a PoC session with the PoC Box, or using other means such as messaging or mail delivery. The UE PoC Box supports the same functionality, although it is a matter of local device implementation how to present it to the user. From the service point of view, the PoC Box may look like any other regular PoC client, with the difference that when a PoC Box is participating into a session, the service must indicate to session participants the fact that they may be effectively being recorded.

9.2.3 Interworking with External P2T Networks

The *External P2T network* is a concept that encompasses any non-OMA compliant service that provides Push-to-Talk-like services to end users. Provided that the external network is able to interwork via a suitable gateway, such external P2T network can look as any other external PoC network, from the service provider point of view. In order to properly support interoperability, the external network must offer fully compliant interfaces POC-4 and IP-1 to the OMA PoC network, possibly through a protocol translating device [1]. POC-4 and IP-1 interfaces build-up the so-called *Network-to-Network Interface* (NNI), which has little differences in concept from the NNI already defined in OMA PoC1.

Figure 9.2 shows a graphical presentation of OMA PoC – P2T interconnection based on NNI, as described by OMA PoC2.

As an alternative to NNI, two service providers may decide to interconnect an OMA PoC service with a remote P2T service using a client-server (master/slave) approach, rather than a server-to-server interface (as NNI is). In such a case, all subscribers (OMA and P2T ones) need to be provisioned as OMA subscribers, and the translating device connects to the OMA service emulating a regular OMA PoC client on behalf of each remote P2T client. Interfaces POC-1 and POC-3 must be offered to the OMA PoC network to achieve interoperability. OMA PoC2 architecture provides further details and use cases about interconnection of OMA PoC2 towards external networks, including both NNI and client-server approaches.

When PoC-P2T interworking is based on OMA NNI (IP-1 and POC-4) the Translation Device is labelled as a *PoC Interworking Function*. If interconnection is based on a client-server interface (POC-1 and POC-3), the Translation Device is the *PoC Interworking Agent* [2].

The ability to connect OMA PoC networks to other P2T systems let operators and service providers leverage their (potentially) existing investments in pre-OMA standard solutions. Furthermore, it extends the reach of OMA PoC to professional and highly specialized niche markets, where trunking technologies have been used for many years (e.g. PMR and TETRA systems). While PoC2 does not have the capability to substitute these technologies when supporting emergency, security or disaster recovery situations, being able to interoperate with such systems may help operators and end users discover and introduce new ways to communicate to a broader communities of users, thus increasing productivity and efficiency of their everyday duties [1].

9.2.4 PoC Crisis Event Handling and QoE

The *PoC Crisis Event Handling* entity is still at an early standardization stage. The idea is that a special client may be able to request setup of sessions dedicated to handle crisis events. In this case, the Crisis Handling Entity is able to request pre-emptive priority for the PoC

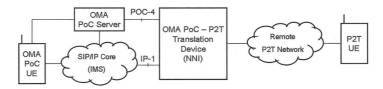

Figure 9.2   OMA PoC2 Network-to-Network interconnection principle

Table 9.1 New OMA PoC2 reference points

Interface	Protocol	Purpose
POC-9 POC-10	SIP RTP MSRP RTCP (MBCP)	These interfaces support the client-server connection between the UE PoC Box and the PoC Service, including support for Voice Talkburst exchange, Discrete Media and/or Media Burst Control messages
POC-11 POC-12	SIP RTP MSRP RTCP (MBCP)	These interfaces support the client-server connection between the NW PoC Box and the PoC service, including support for Voice Talk Burst exchange, Discrete Media and/or Media Burst Control messages
POC-13	XCAP	This interface can be thought as a superset of former POC-5 reference point, as it supports communication between the PoC server and the new Shared XDMS[2]
POC-XX POC-YY	–	Eventually, new interfaces may be defined to support communication between the PoC service and the Crisis Event Handling function

session, it may record sent and received talk bursts, share pictures and video with the participants and establish dynamic group sessions (e.g. set up a crisis handling session with an emergency services team, based on the location of the available participants). This concept must still be developed further in OMA, and will probably be covered in PoCv2.1. As a general idea, the PoC Crisis Event Handling Entity is seen from the network as a PoC client, with special rights and possibly guaranteed network resources and priority, to ensure that it never suffers delay or congestion.

The tight requirements that the Crisis Event handling entity has — in terms of PoC service performance and reliability — lead us to present the *QoE* concept standardized in PoC2. PoC QoE (Quality of Experience) goes a step further beyond traditional QoS (Quality of Service), and defines a suite of levels and performance targets to allocate all PoC users under a service provider within a certain priority level. Depending on user needs and circumstances, the service prioritizes critical or professional users over regular business and residential users. The QoE concept, however, is not linked to any particular architectural element, but on a consistent tailored approach of the whole end-to-end service proposition (thus covering aspects beyond the PoC service, such as network access, user provisioning, IMS priority levels, . . .) [1].

9.2.5 New OMA PoC2 Interfaces

In order to support the new functions described above, OMA has defined a new set of PoC interfaces, not available in the previous release of the enabler. Table 9.1 provides a quick overview of those interfaces and their intended purpose.

---

[2] The OMA XDMv2 compliant Shared XDMS supports a supperset of XCAP applications when compared to XDMv1 Shared XDMS. Hence, POC-13 supports new operations and communication with the new applications hosted at the Shared XDMS.

The rest of elements and interfaces in Figure 9.1 inherit from OMA PoC, XDM and Presence version 1 enablers (except the new XDM concepts described later on). OMA PoC2 additionally defines support for device management functions based on OMA DM, in order to properly configure PoC settings in the UE. Furthermore, OMA PoC2 also supports detailed charging functions, towards an online and/or offline charging system. PoC2 charging is fully based on the same concepts described in 3GPP 32.272 [5] already presented. The main concepts around charging and device management have already been described in chapters 2 and 5. While PoC2 may go into further details about both areas, its main paradigms are aligned with the same functionalities in PoC1.

9.3 OMA XDMv2

OMA XDM is the enabler whose version 2 has earliest reached the *Candidate* status (July 2007). XDMv2 provides a common set of functions that should support other services such as PoC2, Presence 2 and SIMPLE messaging.

As an evolution of XDMv1, XDMv2 supports all requirements defined by OMA for XDMv1 [6], with some architectural changes described here.

The new architecture of OMA XDMv2 is shown in Figure 9.3.

Probably the most important new architectural concepts developed in OMA XDMv2 are: the ability to search for contacts through XDM (incorporating the ability to interconnect XDM services from different operators), and the consolidation of several XDM features and XCAP applications into the new Shared XDMS. These features are presented in deeper detail in the next paragraphs.

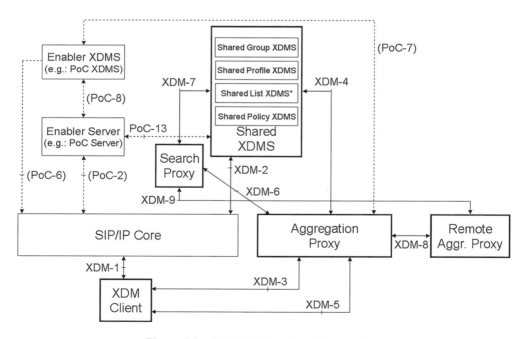

Figure 9.3   OMA XDM version 2 Architecture

- *Search capabilities* are introduced in XDMv2. As an example, XDM should offer a feature similar to those found in existing Messaging services, where users can query the system to look for their colleagues or contacts (using their name and surname, email address, . . .) and, upon successful search, add them to their contact list. The *Search Proxy* is the logical element in charge of search operations in XDMv2. Search operations are not natively supported in XCAP, so an OMA specific mechanism has been developed. As a consequence, client search operations make use of the newly defined XDM-5 reference point to support this feature [7].
- *Interconnection of XDM networks*. XDMv1 had no interconnection support. With the inclusion of XDM-8 reference point, XDM networks may be interconnected. As an example a user will theoretically be able to search for her contacts spread through several XDMv2 enabled networks.
- *Consolidation of XCAP applications into the Shared XDMS*. The XDMv1 Shared XDMS is renamed as the *Shared List XDMS*, in XDMv2. The new Shared XDMS element is now a superset of former Shared XDMS, and encompasses additional capabilities:
  - ○ The *Shared List XDMS* inherits functionality from the Shared XDMS in XDMv1. Thus, it basically stores centralized lists to be used by services such as PoC, IM or Presence.
  - ○ *Shared Group XDMS*. This is one of the most important changes in XDMv2. At some point during standardization work, OMA noticed that many PoC group attributes (anonymity, dial-in vs. dial-out groups) were general enough to apply to many different types of services (VoIP, IM). Additionally, since XML is a structured and extensible language based on usage of namespaces, the advantage of easily consolidating group documents into a shared enabler (the Shared Group XDMS) in a hierarchical and ordered way was identified. When feasible, enablers such as PoC would store groups in the Shared Group XDMS, and only in cases when this node is not capable of storing service-specific logic, an enabler XDMS is defined. From the PoCv1 perspective, the Shared Group XDMS inherits (and extends) functionality from the PoC XDMS.

    Observe the shift from storing PoC groups in the PoC XDMS into having them managed by the Shared Groups XDMS function. This paradigm enables more efficient usage of XML documents, promotes sharing of documents across different enablers and reduces the complexity (and cost) of the specific enabler implementation.
  - ○ The *Shared Policy XDMS* stores policy documents, such as PoC User Access Policy documents or Presence Rules. Hence, it handles all the policy and authorization documents associated to all services from a given subscriber. Observe again the importance of this change. As an example, when a user puts another contact in the PoC blacklist, it seems logical that she will like to hide her Presence information from him as well. With the Shared Policy XDMS such operations may be handled easily. Hence, the Shared Policy XDMS inherits all policy document capabilities from version 1 PoC XDMS and Presence XDMS.
  - ○ The Shared Profile XDMS stores user profile information (name, contact details, address) according to user preferences. This information can later on be retrieved by the Search Proxy, whenever a user looks for a contact.

In addition to the features presented briefly in this section, future XDM enabler releases may define new capabilities such as extended interconnection features (nowadays, only search functions are supported), support document history queries, support different types of XDM users (e.g. administrators, superusers), or define new XDM operations such as the ability to copy or forward XML documents.

9.4 OMA Presence Version 2

OMA Presence version 2 basically enriches and extends the features already supported in version 1. Some use cases such as the dynamic PoC group creation based on user's location are built on top of the location capabilities of the Presence enabler, which were defined already in version 1.

One important goal of Presence v2 is to become a truly horizontal and universal enabler, thus supporting information about virtually any relevant service, such as voice, video calling, SMS, MMS, email, PoC or IM. With Presence v2, users should be able to know whether their contacts are available for PoC, for voice and video calls or just for messaging services.

In its aim to provide as detailed information as possible, there are two interesting use cases which are now introduced by OMA Presence 2, namely:

- Timely notification of the loss of coverage to the Presence server.
- Notification of the roaming status of a Presentity.

These features should help watchers properly decide how and when to communicate with a recipient. However, their accurate implementation is highly dependant on how (and how timely) the underlying network (e.g. IMS, SGSN, GGSN) is able to notify this information to the application layer.

Other interesting features are usage of a Presence tailored *SigComp* extension [8], definition of more detailed interdomain Presence communication. The general topic of supporting multidomain Presence service in an efficient and implementable way is an area of interest and active investigation [9].

From the architectural point of view, there are minor changes in the second version of the OMA Presence enabler. Mainly, a new interface is defined to let Presentities (i.e. end users) know who has subscribed to their Presence information and, eventually, decide –in real-time – whether a new Presence subscription from a given watcher is accepted or not. This feature was not prohibited by Presence v1, as it was based on the well-known *watcherinfo* SIP event package [10], but its inclusion is made explicit in version 2.

However, Presence v2 is yet at an early stage, and we expect significant changes as the standardization process evolves. As an example, at the time of writing the last Presence architecture document does not reflect yet how the changes in XDMv2 affect the evolution (or supression) of Presence XDMS and RLS XDMS.

9.5 OMA SIMPLE Messaging

The last service we will briefly comment in this section is OMA SIMPLE instant messaging. The OMA Messaging Working Group has already standardized an *IMPS* service, based on the *Wireless Village* initiative, which supports basic Presence and instant messaging [11]. The 3GPP IMS concept and interest in SIP/SIMPLE standard solutions for mobile and cellular environments led to the evolution of messaging capabilities towards a SIMPLE messaging enabler, whose architecture is shown in Figure 9.4.

The first version of the SIMPLE IM enabler reached the *Candidate* status in September 2007. Given this timeframe, version 2 of OMA XDM and Presence enablers has been directly adopted by SIMPLE messaging. A lot of concepts already developed in PoC have also been generalized for SIMPLE IM, such as the split between Controlling and Participating functions. This paradigm seems to be a general approach to all SIP based services which can involve a *cascading* implementation (connecting users and servers from several domains).

SIMPLE messaging supports both *pager mode* (one-shot) messaging (based on the SIP MESSAGE transaction) and *session-based* messaging, where SIP is used to establish a messaging session and the *Message Session Relay Protocol* (MSRP) [13] is used to carry actual messages during it. Session based messaging can be thought as the parallel use case for a PoC session, where packetized voice is substituted by text messages delivered on top of MSRP chunks. Furthermore, pager mode messaging can be further divided into *real-time* messaging (the message gets instantly delivered because both the sender and receivers are online simultaneously), and *deferred* messaging (the message is stored at the network while the recipient is offline) [14].

Given the rich set of functionalities and messaging paradigms supported by OMA IM, this enabler is a good basis to support, on the one hand, the evolution of current messaging platforms deployed by cellular operators, such as legacy SMS and MMS technologies. On the other hand, SIMPLE IM is a good basis to develop new communication tools to replicate the Internet community, chat and messaging experiences, on top of SIP-based standardized technology.

Finally, it is worth noting that the next step in messaging standardization at OMA may lead to the development of the *converged IP messaging* concept [15]. This work item is at a rather early stage at OMA. However, if we observe similarities between IM and PoC we can detect that — apart from the fact that user plane traffic is distributed using different underlying protocols (RTP vs. MSRP) — a more general messaging concept may encompass both

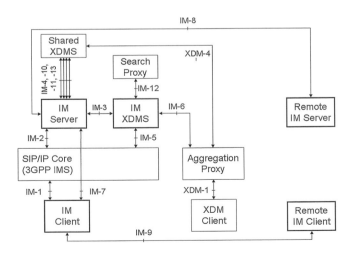

Figure 9.4 OMA SIMPLE IM architecture

services in the future. Hence, the converged IP messaging idea may – although it certainly has to evolve significantly – lead to the definition of a generic architecture supporting many different services, as long as they can be thought as a messaging implementation (e.g. PoC as voice messaging, SMS as text messaging). Thus, the evolution of this work item at OMA will certainly be of interest to the community of operators, service providers and application developers involved in the SIP/SIMPLE arena.

9.6 Summary and Conclusions

In this chapter we have presented an outline of the expected evolution of current specification activities at OMA, covering the evolution of PoC, XDM and Presence enablers. Furthermore, we have briefly covered the status of OMA SIMPLE IM enabler, since it is based on the same set of paradigms and technologies.

An important trend in all cases is the direction towards horizontalizing the final OMA SIP/SIMPLE architecture, with the Shared XDMS taking a predominant role in letting end users manage their group communication settings, profiles and groups at their convenience, from a single modular platform for all SIP based services.

It is important to understand, however, that standardization work is ongoing, and the interested reader should always refer to the latest specification (and, eventually, to the Approved one, as soon as each enabler reaches this status) to confirm or correct the highlights presented in this chapter [16].

9.7 References

[1] OMA: 'Push-to-Talk over Cellular 2 – PoC2 Requirements', Draft Version 2.0; August 2007 (work in progress).

[2] OMA: 'Push-to-Talk over Cellular 2 – PoC2 Architecture', Draft Version 2.0; September 2007 (work in progress).

[3] OMA: 'Push-to-Talk over Cellular 2 – PoC2 User Plane', Draft Version 2.0; September 2007 (work in progress).

[4] G. Camarillo, J. Ott, K. Drage: 'The Binary Floor Control Protocol (BFCP)', RFC 4582, November 2006.

[5] 3GPP TS 32.272 v6.5.0: 'Telecommunication management; Charging management; Push-to-Talk over Cellular (PoC) charging (Release-6)', October 2006.

[6] OMA: 'XML Document Management 2 – XDM Requirements', Candidate Version, July 2007.

[7] OMA: 'XML Document Management 2 – XML Document Management (XDM) Specification', Candidate Version, July 2007.

[8] M. Garci*p1a-Marti*p1n: 'The Presence-Specific Static Dictionary for Signalling Compression (SigComp)', Internet-Draft (work in progress), August 2007.

[9] A. Houri, T. Rang, E. Aoki, et al.: 'Presence Interdomain Scaling Analysis for SIP/SIMPLE', Internet Draft (work in progress), July 2007.

[10] J. Rosenberg: 'A Watcher Information Event Template-Package for the Session Initiation Protocol (SIP)', RFC 3857, August 2004.

[11] OMA: 'Instant Messaging and Presence (IMPS) Enabler Package v1.1', Aproved Enabler, November 2002.

[12] OMA: 'OMA SIMPLE IM v1 – Instant Messaging using SIMPLE Architecture', Candidate Enabler, August 2007.

[13] B. Campbell, R. Mahy, C. Jennings: 'The Message Session Relay Protocol (MSRP)', RFC 4975, September 2007.

[14] OMA: 'OMA SIMPLE IM v1 – Instant Messaging Requirements', Candidate Enabler, August 2007.

[15] OMA 'Converged IP Messaging' Work Item Description (WID0135), May 2006.

[16] OMA Release Program and Specifications: http://www.openmobilealliance.org/release_program/index.html

Index

Abbreviations are explained in a list in the preliminary pages

---